Laboratory Experiments for Organic and Biochemistry

fourth edition

FREDERICK A. BETTELHEIM

JOSEPH M. LANDESBERG

Adelphi University

Harcourt College Publishers

Fort Worth   Philadelphia   San Diego   New York   Orlando   Austin
San Antonio   Toronto   Montreal   London   Sydney   Tokyo

Warnings About Safety Precautions

Some of the experiments contained in this Laboratory Manual involve a degree of risk on the part of the instructor and student. Although performing the experiments is generally safe for the college laboratory, unanticipated and potentially dangerous reactions are possible for a number of reasons, such as improper measuring or handling of chemicals, improper use of laboratory equipment, failure to follow laboratory safety procedures, and other causes. Neither the publisher nor the authors can accept any responsibility for personal injury or property damage resulting from the use of this publication.

Vice President/Publisher: John Vondeling
Vice President/Publisher: Emily Barrosse
Marketing Strategist: Pauline Mula
Developmental Editor: Marc Sherman
Production Manager: Susan Shipe
Art Director: Cara Castiglio, Carol Bleistine
Cover Designer: Karen Gloyd
Cover Credit: Tate Gallery, London/Art Resource, NY. Kandinsky, Wassily. Swinging (Schaukein), 1925.

Laboratory Experiments for Organic and Biochemistry, Fourth Edition
ISBN: 0-03-029204-2

Address for domestic orders:
Harcourt College Publishers, 6277 Sea Harbor Drive, Orlando, FL 32887-6777
1-800-782-4479
e-mail collegesales@harcourt.com

Address for international orders:
International Customer Service, Harcourt, Inc.
6277 Sea Harbor Drive, Orlando FL 32887-6777
(407) 345-3800
Fax (407) 345-4060
e-mail hbintl@harcourt.com

Address for editorial correspondence:
Harcourt College Publishers, Public Ledger Building, Suite 1250,
150 S. Independence Mall West, Philadelphia, PA 19106-3412

Web Site Address
http://www.harcourtcollege.com

Printed in the United States of America
0123456789   202   10 98765432

This book is dedicated to our wives:
Vera S. Bettelheim and Lucy G. Landesberg,
whose help, understanding, and patience
enabled us to write this book.

Preface

In preparing the fourth edition of this Laboratory Manual, we wish to thank our colleagues who made this new edition possible by adopting our Manual for their courses. This fourth edition coincides with the publication of the fourth edition of the textbook: *Introduction to Organic and Biochemistry* by Bettelheim, Brown and March. The textbook shares the outline and the pedagogical philosophy with this book. As in previous editions, we have strived for the clearest possible writing in the procedures. The experiments give the student a meaningful, reliable laboratory experience that consistently work, while covering the basic principles of general, organic and biochemistry. Throughout the years, feedback from different Colleges and Universities made us aware that we have managed to achieve a manual that not only eases the student's task in performing experiments, but also is student friendly. Our new edition maintains this standard and improves upon it.

The major changes in this new edition are as follows: (1) We improved the procedures of all the experiments as a result of our observations of how our students carried out these experiments in our laboratories at Adelphi. (2) Safety issues and waste disposal are reemphasized throughout this edition. (3) We further improved on our aim to minimize the use of hazardous chemicals where possible and to design experiments that work on a semimicro scale. (4) Most Pre-Lab and Post-Lab Questions have been changed or modified.

As in the previous editions, three basic goals were followed in all the experiments: (a) the experiments should illustrate the concepts learned in the classroom; (b) the experiments should be clearly and concisely written so that students will easily understand the task at hand, will work with minimal supervision because the manual provides enough information on experimental procedures, and will be able to perform the experiments in a $2^1/_2$-hr. laboratory period; (c) the experiments should not only be simple demonstrations, but also should contain a sense of discovery.

It did not escape our attention that in adopting this manual of Laboratory Experiments, the instructor must pay attention to budgetary constraints. All experiments in this manual require only inexpensive equipment, if any. A few spectrophotometers and pH meters are necessary in a number of experiments. A few experiments may require more specialized, albeit inexpensive equipment, for example, a few viscometers.

The 27 experiments in this book will provide suitable choice for the instructor to select about 12 experiments for a one semester course. The following are the principal features of this book:

1. The first 12 experiments illustrate the principles of organic chemistry, and the remaining 15 of biochemistry.

2. Each experiment starts out with background information that goes beyond the textbook material. All the relevant principles and their applications are reviewed in this background section.

3. The procedure part provides a step-by-step description of the experiments. Clarity of writing in this section is of utmost importance for successful execution of the experiments. **Caution!** signs alert the students when dealing with dangerous chemicals, such as strong acids or bases.

4. Pre-Lab Questions are provided to familiarize the students with the concepts and procedures before they start the experiments. By requiring the students to answer these questions and by grading their answers, we accomplish the task of preparing the students for the experiments.

5. In the Report Sheet we not only ask for the registration of the raw data, but we also request some calculations to yield secondary data.

6. The Post-Lab Questions are designed so that the student should be able to reflect upon the results, interpret them, and relate their significance.

7. At the end of the book in Appendix 3, we provide the Stockroom Personnel with detailed instructions on preparation of solutions and other chemicals for each experiment. We also give detailed instructions as to how much material is needed for a class of 25 students.

An Instructor's Manual that accompanies this book is **solely for the use of the Instructor.** It helps in the grading process by providing ranges of the experimental results we obtained from class use. In addition it alerts the instructor to some of the difficulties that may be encountered in certain experiments. The disposal of waste material is discussed for each experiment.

We hope that you will find our book of Laboratory Experiments helpful in instructing your students. We anticipate that students will like the book and find it inspiring in studying different aspects of chemistry.

Garden City, NY
April 2000

Frederick A. Bettelheim
Joseph M. Landesberg

Acknowledgments

These experiments have been used by our colleagues over the years and their criticism and expertise were instrumental in refinement of the experiments. We thank Stephen Goldberg, Robert Halliday, Cathy Ireland, Mahadevappa Kumbar, Jerry March, Sung Moon, Donald Opalecky, Reuben Rudman, Charles Shopsis, Kevin Terrance, and Stanley Windwer for their advice and helpful comments. We acknowledge the contributions of Dr. Jessie Lee, Community College of Philadelphia.

We also thank the following reviewers for their thoughtful comments and suggestions: R.E. Bozak, California State University, Hayward; Robert Bruner, Contra Costa College; Bridget Dube, San Antonio College; Katherine Jimison, Cuesta College; Jo Kohn, Olympic College; Margareta Séquin, San Francisco State University; and Mona Wahby, Macomb Community College.

We extend our appreciation to the entire staff at Harcourt College Publishers, especially to John Vondeling, Vice President/Publisher, and Marc Sherman, Developmental Editor, for their encouragement and excellent efforts in producing this book.

Table of Contents

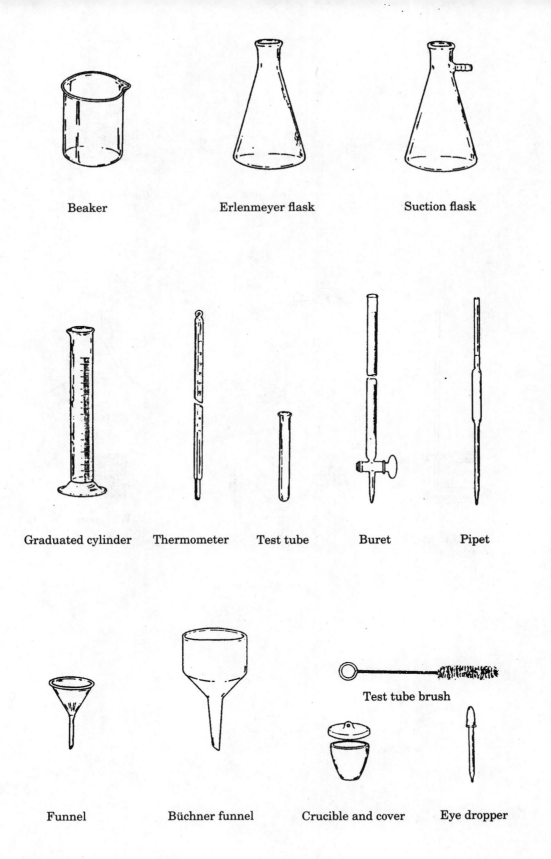

Beaker   Erlenmeyer flask   Suction flask

Graduated cylinder   Thermometer   Test tube   Buret   Pipet

Test tube brush

Funnel   Büchner funnel   Crucible and cover   Eye dropper

Figure 1 • Common laboratory equipment (From Weiner, S.A., and Peters, E.I.: *Introduction to Chemical Principles.* W.B. Saunders, Philadelphia, 1980.)

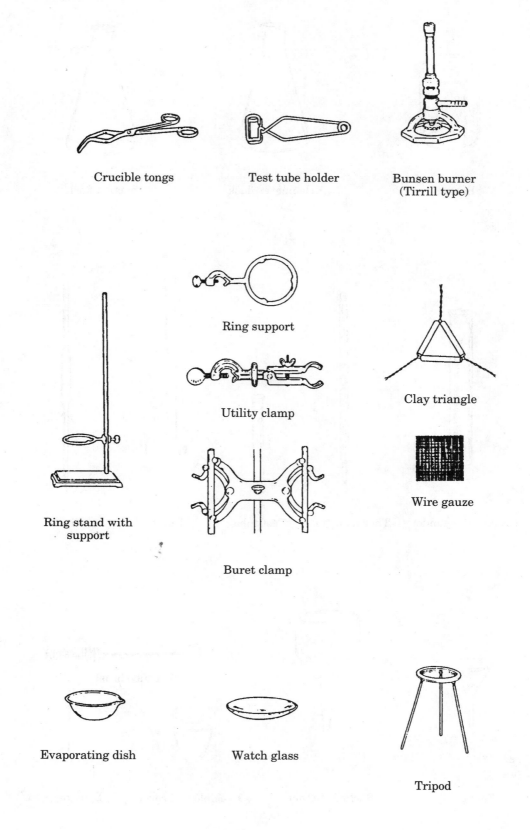

Crucible tongs

Test tube holder

Bunsen burner
(Tirrill type)

Ring support

Utility clamp

Clay triangle

Ring stand with
support

Buret clamp

Wire gauze

Evaporating dish

Watch glass

Tripod

Figure 1 • (continued)

Practice Safe Laboratory

A few precautions can make the laboratory experience relatively hazard-free and safe. These experiments are on a small scale and as such, many of the dangers found in the chemistry laboratory have been minimized. In addition to specific regulations that you may have for your laboratory, the following **DO** and **DON'T RULES** should be observed at all times.

DO RULES

❑ **Do wear approved safety glasses or goggles at all times.**

The first thing you should do after you enter the laboratory is to put on your safety eyewear. The last thing you should do before you leave the laboratory is to remove them. Contact lens wearers must wear additional safety goggles; prescription glasses can be used instead.

❑ **Do wear protective clothing.**

Wear sensible clothing in the laboratory: e.g., no shorts, no tank tops, no sandals. Be covered from the neck to the feet. Laboratory coats or aprons are recommended. Tie back long hair, out of the way of flames.

❑ **Do know the location and use of all safety equipment.**

This includes eyewash facilities, fire extinguishers, fire showers, and fire blankets. In case of fire, do not panic, clear out of the immediate area, and call your instructor for help.

❑ **Do use proper techniques and procedures.**

Closely follow the instructions given in this laboratory manual. These experiments have been student tested; however, accidents do occur but can be avoided if the steps for an experiment are followed. Pay heed to the **Caution!** signs in a procedure.

❑ **Do discard waste material properly.**

Organic chemical waste should be collected in appropriate waste containers and *not flushed down sink drains*. Dilute, nontoxic solutions may be washed down the sink with plenty of water. Insoluble and toxic waste chemicals should be collected in properly labeled waste containers. Follow the directions of your instructor for alternative or special procedures.

❑ **Do be alert, serious, and responsible.**

The best way you can prepare for an experiment is to read the procedure carefully and be aware of the hazards before stepping foot into the laboratory.

DON'T RULES

❑ **Do not eat or drink in the laboratory.**

Consume any food or drink before entering the laboratory. Chemicals could get into food or drinks, causing illness. If you must take a break, wash your hands thoroughly before leaving.

❑ **Do not smoke in the laboratory.**

Smoke only in designated smoking areas outside the laboratory. Flammable gases and volatile flammable reagents could easily explode.

❑ **Do not taste any chemicals or breathe any vapors given off by a reaction.**

If there is a need to smell a chemical, you will be shown how to do it safely.

❑ **Do not get any chemicals on your skin.**

Wash off the exposed area with plenty of water should this happen. Notify your instructor at once. Wear gloves as indicated by your instructor.

❑ **Do not clutter your work area.**

Your laboratory manual and the necessary chemicals, glassware, and hardware are all that should be on your benchtop. This will avoid spilling chemicals and breaking glassware.

❑ **Do not enter the chemical storage area or remove chemicals from the supply area.**

Everyone must have access to the chemicals for the day's experiment. Removal of a chemical from the storage or supply area only complicates the proper execution of the experiment for the other students.

❑ **Do not perform unauthorized experiments.**

Any experiment not authorized presents a hazard to any person in the immediate area.

❑ **Do not take unnecessary risks.**

These **DO** and **DON'T RULES** for a safe laboratory are not an exhaustive list, but are a minimum list of precautions that will make the laboratory a safe and fun activity. Should you have any questions about a hazard, ask your instructor *first*—not your laboratory partner. Finally, if you wish to know about the dangers of any chemical you work with, read the Material Safety Data Sheet (MSDS). These sheets should be on file in the chemistry department office.

Experiment 1

Structure in organic compounds: use of molecular models. I

Background

The study of organic chemistry usually involves those molecules which contain carbon. Thus a convenient definition of *organic chemistry* is the chemistry of carbon compounds.

There are several characteristics of organic compounds that make their study interesting:

a. Carbon forms strong bonds to itself as well as to other elements; the most common elements found in organic compounds, other than carbon, are hydrogen, oxygen, and nitrogen.

b. Carbon atoms are generally tetravalent. This means that carbon atoms in most organic compounds are bound by four covalent bonds to adjacent atoms.

c. Organic molecules are three-dimensional and occupy space. The covalent bonds which carbon makes to adjacent atoms are at discrete angles to each other. Depending on the type of organic compound, the angle may be 180°, 120°, or 109.5°. These angles correspond to compounds which have triple bonds (1), double bonds (2), and single bonds (3), respectively.

$$-C \equiv C- \qquad\qquad {>}C{=}C{<} \qquad\qquad -\overset{|}{\underset{|}{C}}-\overset{|}{\underset{|}{C}}-$$

$$(1) \qquad\qquad\qquad (2) \qquad\qquad\qquad (3)$$

d. Organic compounds can have a limitless variety in composition, shape, and structure.

Thus, while a molecular formula tells the number and type of atoms present in a compound, it tells nothing about the structure. The structural formula is a two-dimensional representation of a molecule and shows the sequence in which the atoms are connected and the bond type. For example, the molecular formula, C_4H_{10}, can be represented by two different structures: butane (4) and 2-methylpropane (isobutane) (5).

$$H-\overset{\overset{\textstyle H}{|}}{\underset{\underset{\textstyle H}{|}}{C}}-\overset{\overset{\textstyle H}{|}}{\underset{\underset{\textstyle H}{|}}{C}}-\overset{\overset{\textstyle H}{|}}{\underset{\underset{\textstyle H}{|}}{C}}-\overset{\overset{\textstyle H}{|}}{\underset{\underset{\textstyle H}{|}}{C}}-H$$

Butane (4)

$$H-\overset{\overset{\textstyle H}{|}}{\underset{\underset{\textstyle H}{|}}{C}}-\overset{\overset{\textstyle H}{|}}{\underset{\underset{\textstyle H}{|}}{C}}-\overset{\overset{\textstyle H}{|}}{\underset{\underset{\textstyle H}{|}}{C}}-H$$

2-Methylpropane (5)
(Isobutane)

Consider also the molecular formula, C_2H_6O. There are two structures which correspond to this formula: dimethyl ether (6) and ethanol (ethyl alcohol) (7).

$$\begin{array}{ccccc} & H & & H & \\ & | & & | & \\ H- & C & -O- & C & -H \\ & | & & | & \\ & H & & H & \end{array} \qquad \begin{array}{ccc} H & H & \\ | & | & \\ H-C & -C & -O-H \\ | & | & \\ H & H & \end{array}$$

Dimethyl ether (6)   Ethanol (7)
(Ethyl alcohol)

In the pairs above, each structural formula represents a different compound. Each compound has its own unique set of physical and chemical properties. Compounds with the same molecular formula but with different structural formulas are called *isomers*.

The three-dimensional character of molecules is expressed by its stereochemistry. By looking at the *stereochemistry* of a molecule, the spatial relationships between atoms on one carbon and the atoms on an adjacent carbon can be examined. Since rotation can occur around carbon-carbon single bonds in open chain molecules, the atoms on adjacent carbons can assume different spatial relationships with respect to each other. The different arrangements that atoms can assume as a result of a rotation about a single bond are called *conformations*. A specific conformation is called a *conformer*. While individual isomers can be isolated, conformers cannot since interconversion, by rotation, is too rapid.

Conformers may be represented by projections through the use of two conventions, as shown in Fig. 1.1. These projections attempt to show on a flat surface how three-dimensional objects, in this case organic molecules, might look in three-dimensional space.

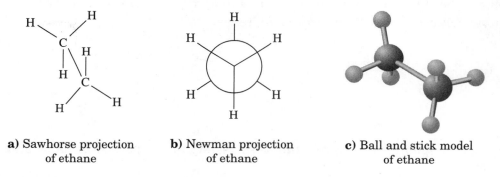

a) Sawhorse projection of ethane   **b)** Newman projection of ethane   **c)** Ball and stick model of ethane

Figure 1.1 • Molecular representations.

The *sawhorse projection* views the carbon-carbon bond at an angle and, by showing all the bonds and atoms, shows their spatial arrangements. The *Newman projection* provides a view along a carbon-carbon bond by sighting directly along the carbon-carbon bond. The near carbon is represented by a circle, and bonds attached to it are represented by lines going to the center of the circle. The carbon behind is not visible (since it is blocked by the near carbon), but the bonds attached to it are partially visible and are represented by lines going to the edge of the circle. With Newman projections, rotations show the spatial relationships of atoms on adjacent carbons easily. Two conformers that represent extremes are shown in Fig. 1.2.

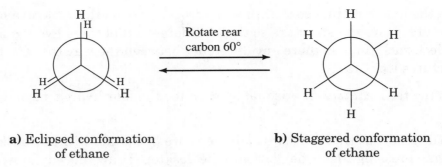

a) Eclipsed conformation
of ethane

b) Staggered conformation
of ethane

Figure 1.2 • Two conformers of ethane.

The *eclipsed* conformation has the bonds (and the atoms) on the adjacent carbons as close as possible. The *staggered* conformation has the bonds (and the atoms) on adjacent carbons as far as possible. One conformation can interconvert into the other by rotation around the carbon-carbon bond axis.

The three-dimensional character of molecular structure is shown through molecular model building. With molecular models, the number and types of bonds between atoms and the spatial arrangements of the atoms can be visualized for the molecules. This allows comparison of isomers and of conformers for a given set of compounds. The models also will let you see what is meant by *chemical equivalence*. Here *equivalence* relates to those positions or to those hydrogens on carbon(s) in an organic molecule that are equal in terms of chemical reactivity. In the case of hydrogen, replacement of any one of the equivalent hydrogens in a molecule by a substituent (any atom or group of atoms, for example, Cl or OH, respectively) leads to the identical substituted molecule.

Objectives

1. To use models to visualize structure in organic molecules.
2. To build and compare isomers having a given molecular formula.
3. To explore the three-dimensional character of organic molecules.
4. To demonstrate equivalence of hydrogens in organic molecules.

Procedure

Obtain a set of ball-and-stick molecular models from the laboratory instructor. The set contains the following parts (other colored spheres may be substituted as available):

- 2 Black spheres representing *Carbon*; this tetracovalent element has four holes;
- 6 Yellow spheres representing *Hydrogen*; this monovalent element has one hole;
- 2 Colored spheres representing the *halogen Chlorine*; this monovalent element has one hole;
- 1 Blue sphere representing *Oxygen*; this divalent element has two holes;
- 8 Sticks to represent bonds.

1. With your models, construct the molecule methane. Methane is a simple hydrocarbon consisting of one carbon and four hydrogens. After you put the model together, answer the questions below in the appropriate space on the Report Sheet.

a. With the model resting so that three hydrogens are on the desk, examine the structure. Move the structure so that a different set of three hydrogens are on the desk each time. Is there any difference between the way that the two structures look (1a)?

b. Does the term *equivalent* adequately describe the four hydrogens of methane (1b)?

c. Tilt the model so that only two hydrogens are in contact with the desk and imagine pressing the model flat onto the desktop. Draw the way in which the methane molecule would look in two-dimensional space (1c). This is the usual way that three-dimensional structures are written.

d. Using a protractor, measure the angle H—C—H on the model (1d).

2. Replace one of the hydrogens of the methane model with a colored sphere, which represents the halogen chlorine. The new model is chloromethane (methyl chloride), CH_3Cl. Position the model so that the three hydrogens are on the desk.

a. Grasp the atom representing chlorine and tilt it to the right, keeping two hydrogens on the desk. Write the structure of the projection on the Report Sheet (2a).

b. Return the model to its original position and then tilt as before, but this time to the left. Write this projection on the Report Sheet (2b).

c. While the projection of the molecule changes, does the structure of chloromethane change (2c)?

3. Now replace a second hydrogen with another chlorine sphere. The new molecule is dichloromethane, CH_2Cl_2.

a. Examine the model as you twist and turn it in space. Are the projections given below isomers of the molecule CH_2Cl_2 or representations of the same structure only seen differently in three dimensions (3a)?

$$
\begin{array}{cccc}
\text{H} & \text{Cl} & \text{H} & \text{Cl} \\
| & | & | & | \\
\text{Cl}-\text{C}-\text{H} & \text{H}-\text{C}-\text{Cl} & \text{Cl}-\text{C}-\text{Cl} & \text{H}-\text{C}-\text{H} \\
| & | & | & | \\
\text{Cl} & \text{H} & \text{H} & \text{Cl}
\end{array}
$$

4. Construct the molecule ethane, C_2H_6. Note that you can make ethane from the methane model by removing a hydrogen and replacing the hydrogen with a methyl group, $-CH_3$.

a. Write the structural formula for ethane (4a).

b. Are all the hydrogens attached to the carbon atoms equivalent (4b)?

c. Draw a sawhorse representation of ethane. Draw a staggered and an eclipsed Newman projection of ethane (4c).

d. Replace any hydrogen in your model with chlorine. Write the structure of the molecule chloroethane (ethyl chloride), C_2H_5Cl (4d).

e. Twist and turn your model. Draw two Newman projections of the chloroethane molecule (4e).

f. Do the projections that you drew represent different isomers or conformers of the same compound (4f)?

5. Dichloroethane, $C_2H_4Cl_2$

a. In your molecule of chloroethane, if you choose to remove another hydrogen note that you now have a choice among the hydrogens. You can either remove a hydrogen from the carbon to which the chlorine is attached, or you can remove a hydrogen from the carbon that has only hydrogens attached. First, remove the hydrogen from the carbon that has the chlorine attached and replace it with a second chlorine. Write its structure on the Report Sheet (5a).

b. Compare this structure to the model which would result from removal of a hydrogen from the other carbon and its replacement by chlorine. Write its structure (5b) and compare it to the previous example. One isomer is 1,1-dichloroethane; the other is 1,2-dichloroethane. Label the structures drawn on the Report Sheet with the correct name.

c. Are all the hydrogens of chloroethane equivalent? Are some of the hydrogens equivalent? Label those hydrogens which are equivalent to each other (5c).

6. Butane

a. Butane has the formula C_4H_{10}. With help from a partner, construct a model of butane by connecting the four carbons in a series (C—C—C—C) and then adding the needed hydrogens. First, orient the model in such a way that the carbons appear as a straight line. Next, tilt the model so that the carbons appear as a zig-zag line. Then, twist around any of the C—C bonds so that a part of the chain is at an angle to the remainder. Draw each of these structures in the space on the Report Sheet (6a). Note that the structures you draw are for the same molecule but represent a different orientation and projection.

b. Sight along the carbon-carbon bond of $\overset{2}{C}$ and $\overset{3}{C}$ on the butane chain: $\overset{1}{C}H_3$—$\overset{2}{C}H_2$—$\overset{3}{C}H_2$—$\overset{4}{C}H_3$. Draw a staggered Newman projection. Rotate the C_2 carbon clockwise by 60°; draw the eclipsed Newman projection. Again, rotate the C_2 carbon clockwise by 60°; draw the Newman projection. Is the last projection staggered or eclipsed (6b)? Continue rotation of the C_2 carbon clockwise by 60° increments and observe the changes that take place.

c. Examine the structure of butane for equivalent hydrogens. In the space on the Report Sheet (6c), redraw the structure of butane and label those hydrogens which are equivalent to each other. On the basis of this examination, predict how many monochlorobutane isomers (C_4H_9Cl) could be obtained from the structure you drew in 6c (6d). Test your prediction by replacement of hydrogen by chlorine on the models. Draw the structures of these isomers (6e).

d. Reconstruct the butane system. First, form a three-carbon chain, then connect the fourth carbon to the center carbon of the three-carbon chain. Add the necessary hydrogens. Draw the structure of 2-methylpropane (isobutane) (6f).

Can any manipulation of the model, by twisting or turning of the model or by rotation of any of the bonds, give you the butane system? If these two, butane and 2-methylpropane (isobutane), are *isomers*, then how may we recognize that any two structures are isomers (6g)?

    e. Examine the structure of 2-methylpropane for equivalent hydrogens. In the space on the Report Sheet (6h), redraw the structure of 2-methylpropane and label the equivalent hydrogens. Predict how many monochloroisomers of 2-methylpropane could be formed (6i) and test your prediction by replacement of hydrogen by chlorine on the model. Draw the structures of these isomers (6j).

7. C_2H_6O

    a. There are two isomers with the molecular formula, C_2H_6O, ethanol (ethyl alcohol) and dimethyl ether. With your partner, construct both of these isomers. Draw these isomers on the Report Sheet (7a) and name each one.

    b. Manipulate each model. Can either be turned into the other by a simple twist or turn (7b)?

    c. For each compound, label those hydrogens which are equivalent. How many sets of equivalent hydrogens are there in ethanol (ethyl alcohol) and dimethyl ether (7c)?

8. Optional: Butenes

    a. If springs are available for the construction of double bonds, construct 2-butene, CH_3—CH=CH—CH_3. There are two isomers for compounds of this formulation: the isomer with the two —CH_3 groups on the same side of the double bond, *cis*-2-butene; and the isomer with the two —CH_3 groups on opposite sides of the double bond, *trans*-2-butene. Draw these two structures on the Report Sheet (8a).

    b. Can you twist, turn, or rotate one model into the other? Explain (8b).

    c. How many bonds are connected to any single carbon of these structures (8c)?

    d. With the protractor, measure the C—C=C angle (8d).

    e. Construct methylpropene, $CH_3-C=CH_2$.
$$\qquad\qquad\qquad\qquad\qquad\ \ |$$
$$\qquad\qquad\qquad\qquad\quad CH_3$$

Can you have a *cis*- or a *trans*- isomer in this system (8e)?

9. Optional: Butynes

    a. If springs are available for the construction of triple bonds, construct 2-butyne, CH_3—C≡C—CH_3. Can you have a *cis*- or a *trans*- isomer in this system (9a)?

    b. With the protractor, measure the C—C≡C angle (9b).

    c. Construct a second butyne with your molecular models and springs. How does this isomer differ from the one in (a) above (9c)?

Chemicals and Equipment

1. Molecular models (you may substitute other available colors for the spheres):
   - 2 Black spheres
   - 6 Yellow spheres
   - 2 Colored spheres (e.g., green)
   - 1 Blue sphere
   - 8 Sticks
2. Protractor
3. Optional: 3 springs

Experiment 2

Stereochemistry: use of molecular models. II

Some molecular variations that acyclic organic molecules can take are:

1. *Constitutional isomerism*. Molecules can have the same molecular formula but different arrangements of atoms.

   a. *skeletal isomerism*: structural isomers where differences are in the order in which atoms that make up the skeleton are connected; e.g., C_4H_{10}

$$CH_3CH_2CH_2CH_3 \qquad \begin{array}{c} CH_3 \\ | \\ CH_3-CH-CH_3 \end{array}$$

   Butane   2-Methylpropane

   b. *positional isomerism*: structural isomers where differences are in the location of a functional group; e.g., C_3H_7Cl

$$CH_3CH_2CH_2-Cl \qquad \begin{array}{c} Cl \\ | \\ CH_3-CH-CH_3 \end{array}$$

   1-Chloropropane   2-Chloropropane

2. *Stereoisomerism*. Molecules which have the same order of attachment of atoms but differ in the arrangement of the atoms in three-dimensional space.

   a. *cis-/trans- isomerism*: molecules that differ due to the geometry of substitution around a double bond; e.g., C_4H_8

$$\begin{array}{c} CH_3 \qquad CH_3 \\ \diagdown \quad \diagup \\ C=C \\ \diagup \quad \diagdown \\ H \qquad\quad H \end{array} \qquad\qquad \begin{array}{c} CH_3 \qquad H \\ \diagdown \quad \diagup \\ C=C \\ \diagup \quad \diagdown \\ H \qquad\quad CH_3 \end{array}$$

   cis-2-Butene   *trans*-2-Butene

   b. *conformational isomerism*: variation in acyclic molecules as a result of a rotation about a single bond; e.g., ethane, CH_3—CH_3

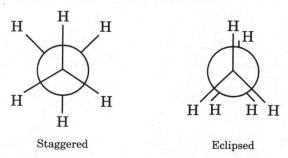

   Staggered   Eclipsed

In this experiment, we will further investigate stereoisomerism by examining a cyclic system, cyclohexane, and several acyclic tetrahedral carbon systems. The latter possess more subtle characteristics as a result of the spatial arrangement of the component atoms. We will do this by building models of representative organic molecules, then studying their properties.

Objectives

1. To use models to study the conformations of cyclohexane.
2. To use models to distinguish between chiral and achiral systems.
3. To define and illustrate enantiomers, diastereomers, and meso forms.
4. To learn how to represent these systems in two-dimensional space.

Procedure

You will build models and then you will be asked questions about the models. You will provide answers to these questions in the appropriate places on the Report Sheet. In doing this laboratory, it will be convenient if you tear out the Report Sheet and keep it by the Procedure as you work through the exercises. In this way, you can answer the questions without unnecessarily turning pages back and forth.

Cyclohexane

Obtain a model set of "atoms" that contain the following:
- 8 Carbon components—model atoms with 4 holes at the tetrahedral angle (e.g., black);
- 2 Substituent components (halogens)—model atoms with 1 hole (e.g., red);
- 18 Hydrogen components—model atoms with 1 hole (optional) (e.g., white);
- 24 Connecting links—bonds.

1. Construct a model of cyclohexane by connecting 6 carbon atoms in a ring; then into each remaining hole insert a connecting link (bond) and, if available, add a hydrogen to each.

   a. Is the ring rigid or flexible, that is, can the ring of atoms move and take various arrangements in space, or is the ring of atoms locked into only one configuration (1a)?

   b. Of the many configurations, which appears best for the ring—a planar or a puckered arrangement (1b)?

   c. Arrange the ring atoms into a *chair* conformation (Fig. 2.1a) and compare it to the picture of the lounge chair (Fig. 2.1b). (Does the term fit the picture?)

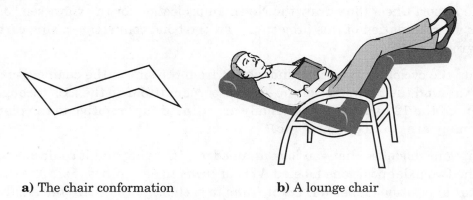

a) The chair conformation   **b)** A lounge chair

Figure 2.1 • The chair conformation for a 6-carbon ring.

2. With the model in the chair conformation, rest it on the tabletop.

   a. How many hydrogens are in contact with the tabletop (2a)?

   b. How many hydrogens point in a direction 180° opposite to these (2b)?

   c. Take your pencil and place it into the center of the ring perpendicular to the table. Now, rotate the ring around the pencil; we'll call this an *axis of rotation*. How many hydrogens are on bonds parallel to this axis (2c)? These hydrogens are called the *axial* hydrogens, and the bonds are called the *axial* bonds.

   d. If you look at the perimeter of the cyclohexane system, the remaining hydrogens lie roughly in a ring perpendicular to the axis through the center of the molecule. How many hydrogens are on bonds lying in this ring (2d)? These hydrogens are called *equatorial* hydrogens, and the bonds are called the *equatorial* bonds.

   e. Compare your model to the diagrams in Fig. 2.2 and be sure you are able to recognize and distinguish between axial and equatorial positions.

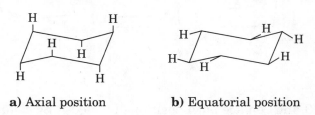

a) Axial position   **b)** Equatorial position

Figure 2.2 • Axial and equatorial hydrogens in the chair conformation.

   In the space provided on the Report Sheet (2e), draw the structure of cyclohexane in the chair conformation with all 12 hydrogens attached. Label all the axial hydrogens, H_a, and all the equatorial hydrogens, H_e. How many hydrogens are labeled H_a (2f)? How many hydrogens are labeled H_e (2g)?

3. Look along any bond connecting any two carbon atoms in the ring. (Rotate the ring and look along a new pair of carbon atoms.) How are the bonds connected to these two carbons arranged? Are they staggered or are they eclipsed (3a)? In the space provided

on the Report Sheet (3b), draw the Newman projection for the view (see Experiment 26 for an explanation of this projection); for the bond connecting a ring carbon, label that group "ring."

4. Pick up the cyclohexane model and view it from the side of the chair. Visualize the "ring" around the perimeter of the system perpendicular to the axis through the center. Of the 12 hydrogens, how many are pointed "up" relative to the plane (4a)? How many are pointed "down" (4b)?

5. Orient your model so that you look at an edge of the ring and it conforms to Fig. 2.3. Are the two axial positions labeled *A cis* or *trans* to each other (5a)? Are the two equatorial positions labeled *B cis* or *trans* to each other (5b)? Are the axial and equatorial positions *A* and *B cis* or *trans* to each other (5c)? Rotate the ring and view new pairs of carbons in the same way. See whether the relationships of positions vary from the above. Position your eye as in Fig. 2.3 and view along the carbon-carbon bond. In the space provided on the Report Sheet (5d), draw the Newman projection. Using this projection, review your answers to 5a, 5b, and 5c.

Figure 2.3
Cyclohexane ring
viewed on edge.

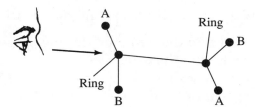

6. Replace one of the axial hydrogens with a colored component atom. Do a "ring flip" by moving one of the carbons *up* and moving the carbon farthest away from it *down* (Fig. 2.4). In what position is the colored component after the ring flip (6a)—axial or equitorial? Do another ring flip. In what position is the colored component now (6b)? Observe all the axial positions and follow them through a ring flip.

Figure 2.4
A "ring flip."

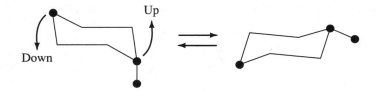

7. Refer to Fig. 2.3 and replace both positions labeled *A* by colored component atoms. Are they *cis* or *trans* (7a)? Do a ring flip. Are the two colored components *cis* or *trans* (7b)? Does the geometry change for the two components as the ring undergoes a ring flip (7c)? Repeat the exercise, replacing atoms in positions labeled *A* and *B* and answer the same three questions for this model.

8. Replace one of the colored components with a methyl, $-CH_3$, group. Manipulate the model so that the $-CH_3$ group is in an axial position; examine the model. Do a ring flip placing the $-CH_3$ in an equatorial position; examine the model. Which of the chair conformations, $-CH_3$ axial or $-CH_3$ equatorial, is more crowded (8a)? What would account for one of the conformations being more crowded than the other (8b)? Which would be of higher energy and thus less stable (8c)? In the space provided on the

Report Sheet (8d), draw the two conformations and connect with equilibrium arrows. Given your answers to 8a, 8b, and 8c, toward which conformation will the equilibrium lay (indicate by drawing one arrow bigger and thicker than the other)?

9. *A substituent group in the equatorial position of a chair conformation is more stable than the same substituent group in the axial position.* Do you agree or disagree? Explain your answer (9).

For the exercises in 10–15, although we will *not* be asking you to draw each and every conformation, we encourage you to practice drawing them in order to gain experience and facility in creating drawings on paper. Your instructor may make these exercises optional.

10. Construct *trans*-1,2-dimethylcyclohexane. By means of ring flips, examine the model with the two —CH_3 groups axial and the two —CH_3 groups equatorial. Which is the more stable conformation? Explain your answer (10).

11. Construct *cis*-1,2-dimethylcyclohexane by placing one —CH_3 group axial and the other equatorial. Do ring flips and examine the two chair conformations. Which is the more stable conformation? Explain your answer (11a). Given the two isomers, *trans*-1,2-dimethylcyclohexane and *cis*-1,2-dimethylcyclohexane, which is the more stable isomer? Explain your answer (11b).

12. Construct *cis*-1,3-dimethylcyclohexane by placing both —CH_3 groups in the axial positions. Do ring flips and examine the two chair conformations. Which is the more stable conformation? Explain your answer (12).

13. Construct *trans*-1,3-dimethylcyclohexane by placing one —CH_3 group axial and the other equatorial. Do ring flips and examine the two chair conformations. Which is the more stable conformation? Explain your answer (13a). Given the two isomers, *trans*-1,3-dimethylcyclohexane and *cis*-1,3-dimethylcyclohexane, which is the more stable isomer? Explain your answer (13b).

14. Construct *trans*-1,4-dimethylcyclohexane by placing both —CH_3 groups axial. Do ring flips and examine the two chair conformations. Which is the more stable conformation? Explain your answer (14).

15. Construct *cis*-1,4-dimethylcyclohexane by placing one —CH_3 group axial and the other equatorial. Do ring flips and examine the two chair conformations. Which is the more stable conformation? Explain your answer (15a). Given the two isomers, *trans*-1,4-dimethylcyclohexane and *cis*-1,4-dimethylcyclohexane, which is the more stable isomer? Explain your answer (15b).

16. Before we leave the cyclohexane ring system, there are some additional ring conformations we can examine. As we move from one cyclohexane chair conformation to another, the *boat* is a transitional conformation between them (Fig. 2.5). Examine a model of the boat conformation by viewing along a carbon-carbon bond, as shown by Fig. 2.5. In the space provided on the Report Sheet (16a), draw the Newman projection for this view and compare with the Newman projection of 5d. By examining the models and comparing the Newman projections, explain which conformation, the chair or the boat, is more stable (16b). Replace the "flagpole" hydrogens by —CH_3 groups. What happens when this is done (16c)? The steric strain can be relieved by twisting the ring and separating the two bulky groups. What results is a *twist boat*.

Figure 2.5
The boat conformation.

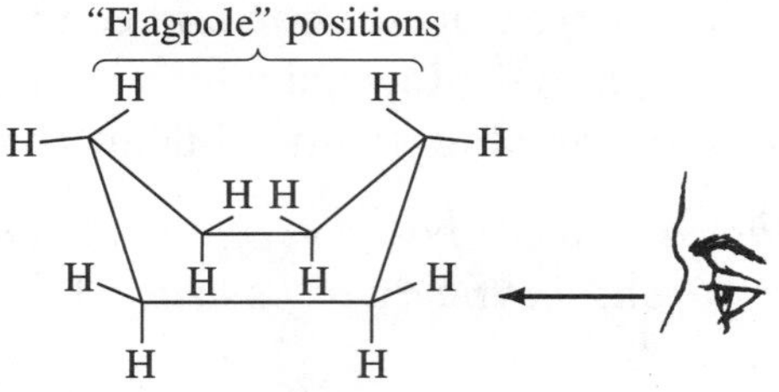

"Flagpole" positions

17. Review the conformations the cyclohexane ring can assume as it moves from one chair conformation to another:

chair ⇌ twist boat ⇌ boat ⇌ twist boat ⇌ chair

Chiral Molecules

For this exercise, obtain a small hand mirror and a model set of "atoms" which contain the following:

- 8 Carbon components—model atoms with four holes at the tetrahedral angle (e.g., black);
- 32 Substituent components—model atoms with one hole in four colors (e.g., 8 red; 8 white; 8 blue; 8 green; or any other colors which your set may have);
- 28 Connecting links—bonds.

Enantiomers

1. Construct a model consisting of a tetrahedral carbon center with four different component atoms attached: red, white, blue, green; each color represents a *different* group or atom attached to carbon. Does this model have a *plane of symmetry* (1a)? A plane of symmetry can be described as a cutting plane—a plane that when passed through a model or object *divides it into two equivalent halves*; the elements on one side of the plane are the exact reflection of the elements on the other side. If you are using a pencil to answer these questions, examine the pencil. Does it have a plane of symmetry (1b)?

2. Molecules without a plane of symmetry are *chiral*. In the model you constructed in no. 1, the tetrahedral carbon is the stereocenter; the molecule is chiral. A simple test for a stereocenter in a molecule is to look for a stereocenter with four different atoms or groups attached to it; this molecule will have no plane of symmetry. On the Report Sheet (2) are three structures; label the stereocenter in each structure with an asterisk (*).

3. Now take the model you constructed in no. 1 and place it in front of a mirror. Construct the model of the image projected in the mirror. You now have two models. If one is the object, what is the other (3a)? Do either have a plane of symmetry (3b)? Are both chiral (3c)? Now try to superimpose one model onto the other, that is, to place one model on top of the other in such a way that all five elements (i.e., the colored atoms) fall exactly one on top of the other. Can you superimpose one model onto the other (3d)? *Enantiomers* are two molecules that are related to each other such that they are *nonsuperimposable mirror images of each other*. Are the two models you have a pair of enantiomers (3e)?

4. Molecules with a plane of symmetry are *achiral*. Replace the blue substituent with a second green one. The model should now have three different substituents attached to the carbon. Does the model now have a plane of symmetry (4a)? Passing the cutting plane through the model, what colored elements does it cut in half (4b)? What is on the left half and right half of the cutting plane (4c)? Place this model in front of the mirror. Construct the model of the image projected in the mirror. You now have two models—an object and its mirror image. Are these two models superimposable on each other (4d)? Are the two models representative of different molecules or identical molecules (4e)?

Each stereoisomer in a pair of enantiomers has the property of being able to rotate monochromatic plane-polarized light. The instrument chemists use to demonstrate this property is called a *polarimeter* (see your text for a further description of the instrument). A pure solution of a single one of the enantiomers (referred to as an *optical isomer*) can rotate the light in either a clockwise (dextrorotatory, +) or a counterclockwise (levorotatory, −) direction. Thus those molecules that are optically active possess a "handedness" or chirality. Achiral molecules are optically inactive and do not rotate the light.

Meso Forms and Diastereomers

5. With your models, construct a pair of enantiomers. From each of the models, remove the same common element (e.g., the white component) and the connecting links (bonds). Reconnect the two central carbons by a bond. What you have constructed is the *meso* form of a molecule, such as *meso*-tartaric acid. How many chiral carbons are there in this compound (5a)?

$$\text{HOOC} - \underset{\underset{\text{OH}}{|}}{\text{C}_a\text{H}} - \underset{\underset{\text{OH}}{|}}{\text{C}_b\text{H}} - \text{COOH}$$

Tartaric acid

Is there a plane of symmetry (5b)? Is the molecule chiral or achiral (5c)?

6. In the space provided on the Report Sheet (6), use circles to indicate the four different groups for carbon C_a and squares to indicate the four different groups for carbon C_b.

7. Project the model into a mirror and construct a model of the mirror image. Are these two models superimposable or nonsuperimposable (7a)? Are the models identical or different (7b)?

8. Now take one of the models you constructed in no. 7, and on one of the carbon centers exchange any two colored component groups. Does the new model have a plane of symmetry (8a)? Is it chiral or achiral (8b)? How many stereocenters are present (8c)? Take this model and one of the models you constructed in no. 7 and see whether they are superimposable. Are the two models superimposable (8d)? Are the two models identical or different (8e)? Are the two models mirror images of each other (8f)? Here we have a pair of molecular models, each with two stereocenters, that are not mirror images of each other. These two examples represent *diastereomers*, stereoisomers that are not related as mirror images.

9. Take the new model you constructed in no. 8 and project it into a mirror. Construct a model of the image in the mirror. Are the two models superimposable (9a)? What term describes the relationship of the two models (9b)?

Thus if we let these three models represent different isomers of tartaric acid, we find that there are three stereoisomers for tartaric acid—a *meso* form and a pair of enantiomers. A *meso* form with any one of the enantiomers of tartaric acid represents a pair of diastereomers. Although it may not be true for this compound because of the *meso* form, in general, if you have **n** stereocenters, there are 2^n stereoisomers possible (see Post-Lab question no. 3).

Drawing Stereoisomers

This section will deal with conventions for representing these three-dimensional systems in two-dimensional space.

10. Construct models of a pair of enantiomers; use tetrahedral carbon and four differently colored components for the four different groups: red, green, blue, white. Hold one of the models in the following way:

a. Grasp the blue group with your fingers and rotate the model until the green and red groups are pointing toward you (Fig. 2.6a). (Use the model which has the green group on the left and the red group on the right.)

b. Holding the model in this way, the blue and white groups point away from you.

c. If we use a drawing that describes a bond pointing toward you as a wedge and a bond pointing away from you as a dashed line, the model can be drawn as shown in Fig. 2.6b.

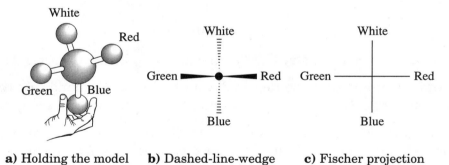

a) Holding the model   **b)** Dashed-line-wedge   **c)** Fischer projection

Figure 2.6 • Projections in two-dimensional space.

If this model were compressed into two-dimensional space, we would get the projection shown in Fig. 2.6c. This is termed a *Fischer projection* and is named after a pioneer in stereochemistry, Emile Fischer. The Fischer projection has the following requirements:

(1) the center of the cross represents the chiral carbon and is in the plane of the paper;

(2) the horizontal line of the cross represents those bonds projecting *out front* from the plane of the paper;

(3) the vertical line of the cross represents bonds projecting *behind* the plane of the paper.

d. In the space provided on the Report Sheet (10), use the enantiomer of the model in Fig. 2.6a and draw both the dashed-line-wedge and Fischer projection.

11. Take the model shown in Fig. 2.6a and rotate by 180° (turn upside down). Draw the Fischer projection (11a). Does this keep the requirements of the Fischer projection (11b)? Is the projection representative of the same system or of a different system (i.e., the enantiomer) (11c)?

In general, if you have a Fischer projection and rotate it in the plane of the paper by 180°, the resulting projection is of the *same* system. Test this assumption by taking the Fischer projection in Fig. 2.6c, rotating it in the plane of the paper by 180°, and comparing it to the drawing you did for no. 11a.

12. Again, take the model shown in Fig. 2.6a. Exchange the red and the green components. Does this exchange give you the enantiomer (12a)? Now exchange the blue and the white components. Does this exchange return you to the original model (12b)?

In general, for a given stereocenter, whether we use the dashed-line wedge or the Fischer projection, an odd-numbered exchange of groups leads to the mirror image of that center; an even-numbered exchange of groups leads back to the original system.

13. Test the above by starting with the Fischer projection given below and carrying out the operations directed in a, b, and c; use the space provided on the Report Sheet (13) for the answers.

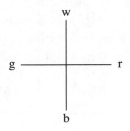

a. Exchange *r* and *g*; draw the Fischer projection you obtain; label this new projection as either the same as the starting model or the enantiomer.

b. Using the new Fischer projection from above, exchange *b* and *w*; draw the Fischer projection you now have.

c. Now rotate the last Fischer projection you obtained by 180°; draw the Fischer projection you now have; label this as either the same as the starting model or the enantiomer.

14. Let us examine models with two stereocenters by using tartaric acid as the example, HOOC—CH(OH)—CH(OH)—COOH; use your colored components to represent the various groups. Hold your models so that each stereoisomer is oriented as in Fig. 2.7. In the space provided on the Report Sheet (14), draw each of the corresponding Fischer projections.

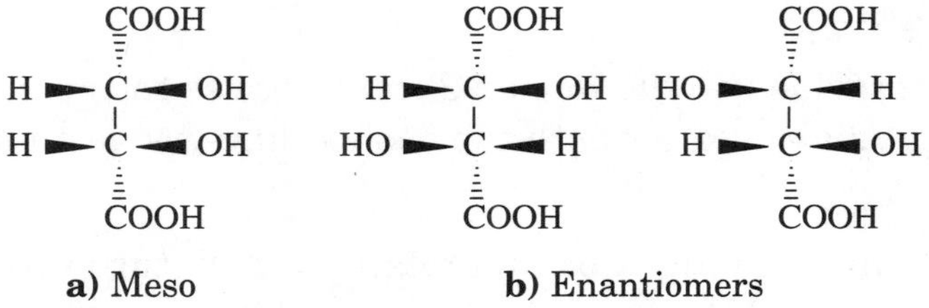

a) Meso   b) Enantiomers

Figure 2.7 • The stereoisomers of tartaric acid.

Circle the Fischer projection that shows a plane of symmetry. Underline all the Fischer projections that would be optically active.

15. Use the Fischer projection of *meso*-tartaric acid and carry out even and odd exchanges of the groups; follow these exchanges with a model. Does an odd exchange lead to an enantiomer, a diastereomer, or to a system identical to the *meso* form (15a)? Does an even exchange lead to an enantiomer, a diastereomer, or to a system identical to the *meso* form (15b)?

Chemicals and Equipment

Model kits vary in size and color of components. Use what is available; other colors may be substituted.

1. Cyclohexane model kit: 8 carbons (black, 4 holes); 18 hydrogens (white, 1 hole); 2 substituents (red, 1 hole); 24 bonds.
2. Chiral model kit: 8 carbons (black, 4 holes); 32 substituents (8 red, 1 hole; 8 white, 1 hole; 8 blue, 1 hole; 8 green, 1 hole); 28 bonds.
3. Hand mirror

Experiment 2

PRE-LAB QUESTIONS

1. What is the most stable conformation for cyclohexane?

2. Look at your hands and your feet. Which term best explains the relationship of the two hands and of the two feet: identical, constitutional, conformational, or enantiomers?

3. What term describes molecules without a plane of symmetry?

4. Label the chiral carbons in the molecules below with an asterisk (*).

5. Those molecules that are optically active possess _____.

NAME SECTION DATE

PARTNER GRADE

Experiment 2

REPORT SHEET

Cyclohexane

1. a.

b.

2. a.

b.

c.

d.

e.

f.

g.

3. a.

b.

4. a.

b.

5. a.

b.

c.

d.

6. a.

   b.

7.   *Trial 1*   *Trial 2*

   a.

   b.

   c.

8. a.

   b.

   c.

   d.

9.

10. e,e or a,a

11. a. a,e or e,a

   b.

12. a,a or e,e

13. a. a,e or e,a

   b.

14. a,a or e,e

15. a. a,e or e,a

   b.

16. a.

   b.

   c.

Enantiomers

1. **a.**

   b.

2.
$$CH_3-\underset{\underset{\textstyle OH}{|}}{CH}-CH_2CH_3 \qquad CH_3-\underset{\underset{\textstyle OH}{|}}{CH}-COOH \qquad ClCH_2-\underset{\underset{\textstyle Br}{|}}{CH}-CH_3$$

3. **a.**

   b.

   c.

   d.

   e.

4. **a.**

   b.

   c.

   d.

   e.

Meso *forms and diastereomers*

5. **a.**

b.

c.

6. $HOOC-\underset{\underset{HO}{|}}{\overset{\overset{H}{|}}{C_a}}-\underset{\underset{OH}{|}}{\overset{\overset{H}{|}}{C_b}}-COOH$ $\qquad$ $HOOC-\underset{\underset{HO}{|}}{\overset{\overset{H}{|}}{C_a}}-\underset{\underset{OH}{|}}{\overset{\overset{H}{|}}{C_b}}-COOH$

7. a.

b.

8. a.

b.

c.

d.

e.

f.

9. a.

b.

Drawing stereoisomers

10.

11. a.

b.

c.

12. a.

b.

13. a.

b.

c.

14.

15. a.

b.

POST-LAB QUESTIONS

1. Which position is more stable for the methyl group in methylcyclohexane: an equatorial position or an axial position? Explain your answer.

2. Draw the Fischer projections for the pair of enantiomers of lactic acid, CH_3—$CH(OH)$—$COOH$.

3. For 2,3-dibromopentane:

a. How many stereoisomers are possible for this compound?

$$
\begin{array}{cc}
Br & Br \\
| & | \\
CH_3-CH-CH-CH_2CH_3
\end{array}
$$

b. Draw Fischer projections for each stereoisomer; label enantiomers. Label any *meso* isomers (if there are any).

4. Determine the relationship between the following pairs of structures: identical, enantiomers, diastereomers.

a.

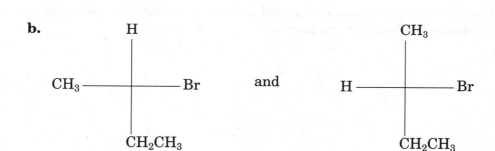

and

b.

$$
\begin{array}{ccc}
& H & \\
& | & \\
CH_3 - & - & Br \\
& | & \\
& CH_2CH_3 &
\end{array}
\qquad \text{and} \qquad
\begin{array}{ccc}
& CH_3 & \\
& | & \\
H - & - & Br \\
& | & \\
& CH_2CH_3 &
\end{array}
$$

Identification of hydrocarbons

Background

The number of known organic compounds totals into the millions. Of these compounds, the simplest types are those which contain only hydrogen and carbon atoms. These are known as *hydrocarbons*. Because of the number and variety of hydrocarbons that can exist, some means of classification is necessary.

One means of classification depends on the way in which carbon atoms are connected. *Chain* aliphatic hydrocarbons are compounds consisting of carbons linked either in a single chain or in a branched chain. *Cyclic* hydrocarbons are aliphatic compounds that have carbon atoms linked in a closed polygon (also referred to as a *ring*). For example, hexane (single) and 2-methylpentane (branched) are chain aliphatic molecules, while cyclohexane is a cyclic aliphatic compound.

$$CH_3CHCH_2CH_2CH_3$$
$$|$$
$$CH_3$$

$$CH_3CH_2CH_2CH_2CH_2CH_3$$

Hexane

2-Methylpentane

Cyclohexane

Another means of classification depends on the type of bonding that exists between carbons. Hydrocarbons which contain only carbon-to-carbon single bonds are called *alkanes*. These are also referred to as *saturated* molecules. Hydrocarbons containing at least one carbon-to-carbon double bond are called *alkenes*, and those compounds with at least one carbon-to-carbon triple bond are called *alkynes*. These are compounds that are referred to as *unsaturated* molecules. Finally, a class of cyclic hydrocarbons that contain a closed loop (sextet) of electrons are called *aromatic* (see Chapter 14 in your text for further details). Table 3.1 distinguishes between the families of hydrocarbons.

With so many compounds possible, identification of the bond type is an important step in establishing the molecular structure. Quick, simple tests on small samples can establish the physical and chemical properties of the compounds by class.

Some of the observed physical properties of hydrocarbons result from the nonpolar character of the compounds. In general, hydrocarbons do not mix with polar solvents such as water or ethyl alcohol. On the other hand, hydrocarbons mix with relatively nonpolar solvents such as ligroin (a mixture of alkanes), carbon tetrachloride, or dichloromethane. Since the density of most hydrocarbons is less than that of water, they will float. Crude oil and crude oil products (home heating oil and gasoline) are mixtures of hydrocarbons; these substances, when spilled on water, spread quickly along the surface because they are insoluble in water.

Table 3.1 Types of Hydrocarbons

Class	Characteristic Bond Type		Example	
I. Aliphatic				
1. Alkane*	$-\overset{\mid}{\underset{\mid}{C}}-\overset{\mid}{\underset{\mid}{C}}-$	single	$CH_3CH_2CH_2CH_2CH_2CH_3$	hexane
2. Alkene†	$\overset{\diagup}{\diagdown}C=C\overset{\diagdown}{\diagup}$	double	$CH_3CH_2CH_2CH_2CH=CH_2$	1-hexene
3. Alkyne†	$-C\equiv C-$	triple	$CH_3CH_2CH_2CH_2C\equiv CH$	1-hexyne
II. Cyclic				
1. Cycloalkane*	$-\overset{\mid}{\underset{\mid}{C}}-\overset{\mid}{\underset{\mid}{C}}-$	single		cyclohexane
2. Cycloalkene†	$\overset{\diagup}{\diagdown}C=C\overset{\diagdown}{\diagup}$	double		cyclohexene
3. Aromatic				benzene
				toluene

*Saturated   †Unsaturated

The chemical reactivity of hydrocarbons is determined by the type of bond in the compound. Although saturated hydrocarbons (alkanes) will burn (undergo *combustion*), they are generally unreactive to most reagents. (Alkanes do undergo a substitution reaction with halogens but require ultraviolet light.) Unsaturated hydrocarbons, alkenes and alkynes, not only burn, but also react by *addition* of reagents to the double or triple bonds. The addition products become saturated, with fragments of the reagent becoming attached to the carbons of the multiple bond. Aromatic compounds, with a higher carbon-to-hydrogen ratio than nonaromatic compounds, burn with a sooty flame as a result of unburned carbon particles being present. These compounds undergo *substitution* in the presence of catalysts rather than an addition reaction.

1. *Combustion*. The major component in "natural gas" is the hydrocarbon methane. Other hydrocarbons used for heating or cooking purposes are propane and butane. The products from combustion are carbon dioxide and water (heat is evolved, also).

$$CH_4 + 2O_2 \longrightarrow CO_2 + 2H_2O$$

$$CH_3CH_2CH_3 + 5O_3 \longrightarrow 3CO_2 + 4H_2O$$

2. *Reaction with bromine.* Unsaturated hydrocarbons react rapidly with bromine in a solution of carbon tetrachloride or cyclohexane. The reaction is the addition of the elements of bromine to the carbons of the multiple bonds.

Alkene

$$CH_3CH{=}CHCH_3 + \underset{\text{Red}}{Br_2} \longrightarrow \underset{\text{Colorless}}{CH_3\underset{}{\overset{\overset{\textstyle Br}{|}}{C}H}{-}\overset{\overset{\textstyle Br}{|}}{C}HCH_3}$$

Alkyne

$$CH_3C{\equiv}CCH_3 + \underset{\text{Red}}{2Br_2} \longrightarrow \underset{\text{Colorless}}{CH_3\overset{\overset{\textstyle Br}{|}}{\underset{\underset{\textstyle Br}{|}}{C}}{-}\overset{\overset{\textstyle Br}{|}}{\underset{\underset{\textstyle Br}{|}}{C}}CH_3}$$

The bromine solution is red; the product that has the bromine atoms attached to carbon is colorless. Thus a reaction has taken place when there is a loss of color from the bromine solution and a colorless solution remains. Since alkanes have only single C—C bonds present, no reaction with bromine is observed; the red color of the reagent would persist when added. Aromatic compounds resist addition reactions because of their "aromaticity": *the possession of a closed loop (sextet) of electrons.* These compounds react with bromine in the presence of a catalyst such as iron filings or aluminum chloride.

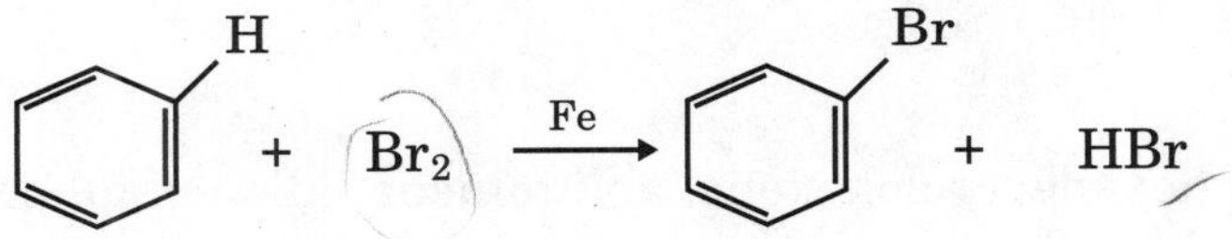

Note that a substitution reaction has taken place and the gas HBr is produced.

3. *Reaction with concentrated sulfuric acid.* Alkenes react with cold concentrated sulfuric acid by addition. Alkyl sulfonic acids form as products and are soluble in H_2SO_4.

$$CH_3{-}CH{=}CH{-}CH_3 + \underset{(H_2SO_4)}{HOSO_2OH} \longrightarrow CH_3{-}\underset{\underset{\textstyle H}{|}}{CH}{-}\underset{\underset{\textstyle OSO_2OH}{|}}{CH}{-}CH_3$$

Saturated hydrocarbons are unreactive (additions are not possible); alkynes react slowly and require a catalyst ($HgSO_4$); aromatic compounds also are unreactive since addition reactions are difficult.

4. *Reaction with potassium permanganate.* Dilute or alkaline solutions of $KMnO_4$ oxidize unsaturated compounds. Alkanes and aromatic compounds are generally unreactive. Evidence that a reaction has occurred is observed by the loss of the purple color of $KMnO_4$ and the formation of the brown precipitate manganese dioxide, MnO_2.

$$3CH_3{-}CH{=}CH{-}CH_3 + \underset{\text{Purple}}{2KMnO_4} + 4H_2O \longrightarrow 3CH_3{-}\underset{\underset{\textstyle OH}{|}}{CH}{-}\underset{\underset{\textstyle OH}{|}}{CH}{-}CH_3 + \underset{\text{Brown}}{2MnO_2} + 2KOH$$

Note that the product formed from an alkene is a glycol.

Objectives

1. To investigate the physical properties, solubility and density, of some hydrocarbons.
2. To compare the chemical reactivity of an alkane, an alkene, and an aromatic compound.
3. To use physical and chemical properties to identify an unknown.

Procedure

CAUTION!

Assume the organic compounds are highly flammable. Use only small quantities. Keep away from open flames. Assume the organic compounds are toxic and can be absorbed through the skin. Avoid contact; wash if any chemical spills on your person. Handle concentrated sulfuric acid carefully. Flush with water if any spills on your person. Potassium permanganate and bromine are toxic; bromine solutions are also corrosive. Although the solutions are dilute, they may cause burns to the skin. Wear gloves when working with these chemicals.

General Instructions

1. The hydrocarbons hexane, cyclohexene, and toluene (alkane, alkene, and aromatic) are available in dropper bottles.

2. The reagents 1% Br_2 in cyclohexane, 1% aqueous $KMnO_4$, and concentrated H_2SO_4 are available in dropper bottles.

3. Unknowns are in dropper bottles labeled A, B, and C. They may include an alkane, an alkene, or an aromatic compound.

4. Record all data and observations in the appropriate places on the Report Sheet.

5. Dispose of all organic wastes as directed by the instructor. *Do not pour into the sink!*

Physical Properties of Hydrocarbons

1. A test tube of 100 × 13 mm will be suitable for this test. When mixing the components, grip the test tube between thumb and forefinger; it should be held firmly enough to keep from slipping but loosely enough so that when the third and fourth fingers tap it, the contents will be agitated enough to mix.

2. *Water solubility of hydrocarbons.* Label six test tubes with the name of the substance to be tested. Place into each test tube 5 drops of the appropriate hydrocarbon: hexane, cyclohexene, toluene, unknown A, unknown B, unknown C. Add about 5 drops of water dropwise into each test tube. Is there any separation of components? Which component is on the bottom; which component is on the top? Mix the contents as described above.

What happens when the contents are allowed to settle? What do you conclude about the density of the hydrocarbon? Is the hydrocarbon *more* dense than water or *less* dense than water? Record your observations. Save these solutions for comparison with the next part.

3. *Solubility of hydrocarbons in ligroin.* Label six test tubes with the name of the substance to be tested. Place into each test tube 5 drops of the appropriate hydrocarbon: hexane, cyclohexene, toluene, unknown A, unknown B, unknown C. Add about 5 drops of ligroin dropwise into each test tube. Is there a separation of components? Is there a bottom layer and top layer? Mix the contents as described above. Is there any change in the appearance of the contents before and after mixing? Compare these test tubes to those from the previous part. Record your observations. Can you make any conclusion about the density of the hydrocarbons from what you actually see?

Chemical Properties of Hydrocarbons

1. *Combustion.* The instructor will demonstrate this test in the fume hood. Place 5 drops of each hydrocarbon and unknown on separate watch glasses. Carefully ignite each sample with a match. Observe the flame and color of the smoke for each of the samples. Record your observations on the Report Sheet.

2. *Reaction with bromine.* Label six clean, dry test tubes with the name of the substance to be tested. Place into each test tube 5 drops of the appropriate hydrocarbon: hexane, cyclohexene, toluene, unknown A, unknown B, unknown C. Carefully add (dropwise and with shaking) 1% Br_2 in cyclohexane. Keep count of the number of drops needed to have the color persist; do not add more than 10 drops. Record your observations. To any sample that gives a negative test after adding 10 drops of bromine solution (i.e., the red color persists), add 5 more drops of 1% Br_2 solution and a small quantity of iron filings; shake the mixture. Place a piece of moistened blue litmus paper on the test tube opening. Record any change in the color of the solution and the litmus paper.

CAUTION!

Use 1% Br_2 solution in the hood; wear gloves when using this chemical.

---

3. *Reaction with $KMnO_4$.* Label six clean, dry test tubes with the name of the substance to be tested. Place into each test tube 5 drops of the appropriate hydrocarbon: hexane, cyclohexene, toluene, unknown A, unknown B, unknown C. Carefully add (dropwise) 1% aqueous $KMnO_4$ solution; after each drop, shake to mix the solutions. Keep count of the number of drops needed to have the color of the permanganate solution persist; do not add more than 10 drops. Record your observations.

4. *Reaction with concentrated H_2SO_4.* Label six clean, dry test tubes with the name of the substance to be tested. Place into each test tube 5 drops of the appropriate hydrocarbon: hexane, cyclohexene, toluene, unknown A, unknown B, unknown C. Place all of the test tubes in an ice bath. *Wear gloves and carefully* add (with shaking) 3 drops of cold, concentrated sulfuric acid to each test tube. Note whether heat is evolved by

feeling the test tube. Note whether the solution has become homogeneous or whether a color is produced. (The evolution of heat or the formation of a homogeneous solution or the appearance of a color is evidence that a reaction has occurred.) Record your observations.

5. *Unknowns*. By comparing the observations you made for your unknowns with that of the known hydrocarbons, you can identify unknowns A, B, and C. Record their identities on your Report Sheet.

Chemicals and Equipment

1. 1% aqueous $KMnO_4$
2. 1% Br_2 in cyclohexane
3. Blue litmus paper
4. Concentrated H_2SO_4
5. Cyclohexene
6. Hexane
7. Iron filings or powder
8. Test tubes
9. Ligroin
10. Toluene
11. Unknowns A, B, and C
12. Watch glasses
13. Ice

Column and paper chromatography: separation of plant pigments

Chromatography is a widely used experimental technique by which a mixture of compounds can be separated into its individual components. Two kinds of chromatographic experiments will be explored. In column chromatography, a mixture of components dissolved in a solvent is poured over a column of solid adsorbent and is eluted with the same or a different solvent. This is therefore a solid-liquid system; the stationary phase (the adsorbent) is solid and the mobile phase (the eluent) is liquid. In paper chromatography, the paper adsorbs water from the atmosphere of the developing chromatogram. (The water is present in the air as vapor, and it may be supplied as one component in the eluting solution.) The water is the stationary phase. The (other) component of the eluting solvent is the mobile phase and carries with it the components of the mixture. This is a liquid-liquid system.

Column chromatography is used most conveniently for preparative purposes, when one deals with a relatively large amount of the mixture and the components need to be isolated in milligrams or grams quantities. Paper chromatography, on the other hand, is used mostly for analytical purposes. Microgram or even picogram quantities can be separated by this technique, and they can be characterized by their R_f number. This number is an index of how far a certain spot moved on the paper.

$$R_f = \frac{\text{Distance of the center of the sample spot from the origin}}{\text{Distance of the solvent front from the origin}}$$

For example, in Fig. 4.1 the R_f values are as follows:

R_f (substance 1) = 3.1 cm/11.2 cm = 0.28 and

R_f (substance 2) = 8.5 cm/11.2 cm = 0.76

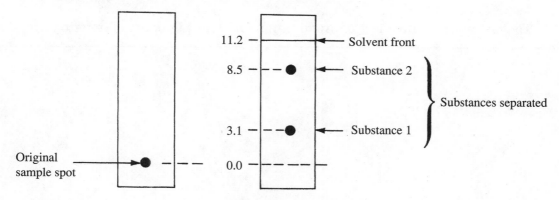

Figure 4.1 • Illustration of chromatograms before and after elution.

Using the R_f values, one is able to identify the components of the mixture with the individual components. The two main pigment components of tomato paste are β-carotene (yellow-orange) and lycopene (red) pigments. Their structures are given below:

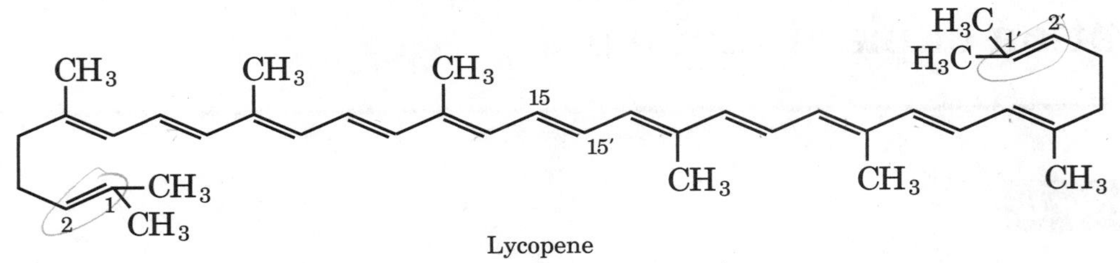

Lycopene

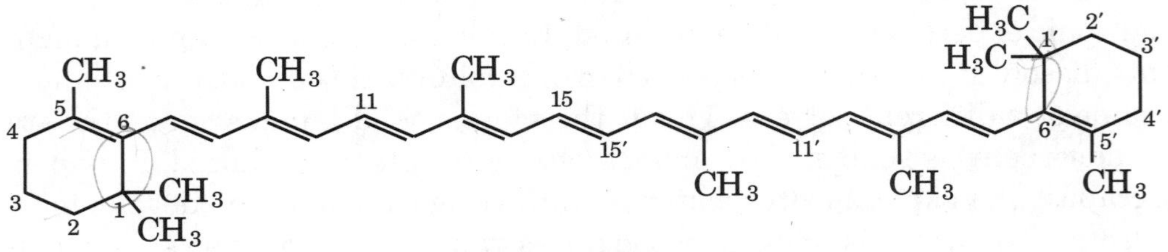

β-Carotene

The colors of these pigments are due to the numerous double bonds in their structure. When bromine is added to double bonds, it saturates them and the color changes accordingly. In the tomato juice "rainbow" experiment, we stir bromine water into the tomato juice. The slow stirring allows the bromine water to penetrate deeper and deeper into the cylinder in which the tomato juice was placed. As the bromine penetrates, more and more double bonds will be saturated. Therefore, you may be able to observe a continuous change, a "rainbow" of colors, starting with the reddish tomato color at the bottom of the cylinder where no reaction occurred (since the bromine did not reach the bottom). Lighter colors will be observed on the top of the cylinder where most of the double bonds have been saturated.

Objectives

1. To compare separation of components of a mixture by two different techniques.
2. To demonstrate the effect of bromination on plant pigments of tomato juice.

Procedure

Paper Chromatography

ethanol
—denatured

30.141 g
4/2.08

11
7
18

1. Obtain a sheet of Whatman no.1 filter paper, cut to size.

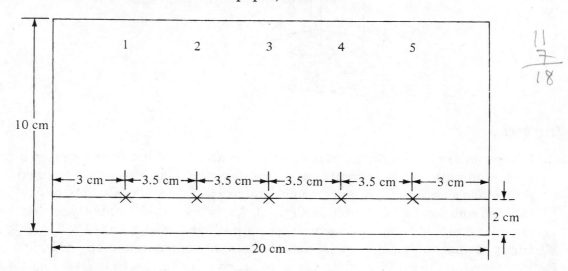

Figure 4.2 • Preparation of chromatographic paper for spotting.

2. Plan the **spotting** of the samples as illustrated on Fig. 4.2. Five spots will be applied. The first and fifth spots will be β-carotene solutions supplied by your instructor. The second, third, and fourth spots will have your tomato paste extracts in different concentrations. Use a pencil to mark the placement of the spots lightly according to Fig. 4.2.

3. Pigments of tomato paste will be **extracted** in two steps.

   (a) Weigh about 10 g of tomato paste in a 50-mL beaker. Add 15 mL of 95% ethanol. Stir the mixture vigorously with a spatula until the paste will not stick to the stirrer. Place a small amount of glass wool (the size of a pea) in a small funnel, blocking the funnel exit. Place the funnel into a 50-mL Erlenmeyer flask and pour the tomato paste–ethanol mixture into the funnel. When the filtration is completed, squeeze the glass wool lightly with your spatula. In this step, we removed the water from the tomato paste and the aqueous components are in the filtrate, which we discard. The residue in the glass wool will be used to extract the pigments.

   (b) Place the residue from the glass wool in a 50-mL beaker. Add 10 mL petroleum ether and stir the mixture for about 2 min. to extract the pigments. Filter the extract as before through a new funnel with glass wool blocking the exit into a new and clean 50-mL beaker. Place the beaker under the hood on a hot plate (or water bath). No open flame, such as a Bunsen burner, is allowed. Evaporate the solvent to about 1 mL volume. Use low heat and take care not to evaporate all the solvent. After evaporation, cover the beaker with aluminum foil.

Figure 4.3
Withdrawing samples
with a capillary tube.

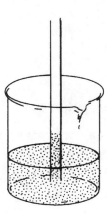

Spotting

4. Place your chromatographic paper on a clean area (another filter paper) in order not to contaminate it. Use separate capillaries for your tomato paste extract and for the β-carotene solution. First, apply your capillary to the extracted pigment by dipping it into the solution as illustrated in Fig. 4.3. Apply the capillary lightly to the chromatographic paper by touching sequentially the spots marked 2, 3, and 4. Make sure you apply only small spots, not larger than 2 mm diameter, by quickly withdrawing the capillary from the paper each time you touch it. (See Fig. 4.4.)

Figure 4.4
Spotting.

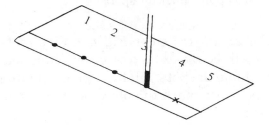

While allowing the spots to dry, use your second capillary to apply spots of β-carotene in lanes 1 and 5. Return to the first capillary and apply another spot of the extract on top of the spots of lanes 3 and 4. Let them dry (Fig. 4.5). Finally, apply one more spot on top of lane 4. Let the spots dry. The unused extract in your beaker should be covered with aluminum foil. Place it in your drawer in the dark to save it for the second part of your experiment.

Figure 4.5
Drying chromatographic spots.

Harcourt, Inc.

Developing the paper chromatogram

5. Curve the paper into a cylinder and staple the edges above the 2-cm line, as is shown in Fig. 4.6.

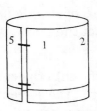

Figure 4.6 • Stapling.

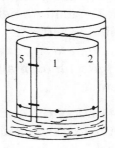

Figure 4.7 • Developing the chromatogram.

6. Pour 20 mL of the eluting solvent (petroleum ether : toluene : acetone in 45 : 1 : 5 ratio, supplied by your instructor) into a 600-mL beaker.

7. Place the stapled chromatogram into the 600-mL beaker, the spots being at the bottom near the solvent surface but **not covered by it**. Cover the beaker with aluminum foil (Fig. 4.7). Allow the solvent front to migrate up to 0.5–1 cm below the edge of the paper. This may take from 15 min. to 1 hr. Make certain by frequent inspection that the **solvent front does not run over the edge of the paper**. Remove the chromatogram from the beaker when the solvent front reaches 0.5–1 cm from the edge; then **proceed to step 11**.

Column Chromatography

8. While you are waiting for the paper chromatogram to develop (step no. 7), you can perform the column chromatography experiment. Take a 25-mL buret. (You may use a chromatographic column, if available, of 1.6 cm diameter and about 13 cm long; see Fig. 4.8. If you use the column instead of the buret, all subsequent quantities below should be doubled.)

Figure 4.8
Chromatographic column.

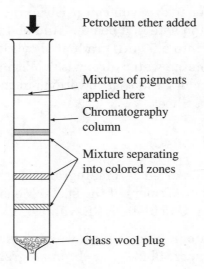

Petroleum ether added

Mixture of pigments applied here

Chromatography column

Mixture separating into colored zones

Glass wool plug

Add a small piece of glass wool and with the aid of a glass rod push it down near the stopcock. Add 15–16 mL of petroleum ether to the buret. Open the stopcock slowly and allow the solvent to fill the tip of the buret. Close the stopcock. You should have 12–13 mL of solvent above the glass wool. Weigh 20 g of aluminum oxide (alumina) in a 100-mL beaker. Place a small funnel on top of your buret. Pour the alumina into the buret. Allow the alumina to settle in order to form a 20-cm column. *Drain* the solvent but *do not allow the column to run dry. Always have at least 0.5 mL of clear solvent above the alumina in the column.* If alumina adheres to the walls of the buret, wash it down with more solvent.

9. Transfer by pipet 0.5–1 mL of the extract you stored in your drawer onto the column. The pipet containing the extract should be placed near the surface of the solvent on top of the column. Touching the walls of the buret with the tip of the pipet, allow the extract to drain slowly on top of the column. Open the stopcock slightly. Allow the sample to enter the column, *but make sure there is a small amount of solvent above the alumina in the column. (The column should never run dry.)* Add 10 or more mL of petroleum ether and wash the sample into the column by opening the stopcock and collecting the eluted solvent in a beaker.

10. As the solvent elutes the sample, you observe the migration of the pigments and their separation into at least two bands. When the fastest-moving pigment band reaches near the bottom of the column, close the stopcock and observe the color of the pigment bands and how far they migrated from the top of the column. Record your observation on the Report Sheet. This concludes the column chromatographic part of the experiment. Discard your solvent in a bottle supplied by your instructor for a later redistillation.

11. Meanwhile your paper chromatogram has developed. You must remove the filter paper from the 600-mL beaker before the solvent front reaches the edges of the paper. *Mark the position of the solvent front with a pencil.* Put the paper standing on its edges under the hood and let it dry.

Tomato Juice "Rainbow"

12. While waiting for the paper to dry, you can perform the following short experiment. Weigh about 15 g of tomato paste in a beaker. Add about 30 mL of water and stir. Transfer the tomato juice into a 50-mL graduated cylinder and, with the aid of a pipet, add 5 mL of saturated bromine water (dropwise). With a glass rod, stir the solution very gently. Observe the colors and their positions in the cylinder. Record your observations on the Report Sheet.

Paper Chromatography (continued)

13. Remove the staples from the dried chromatogram. Mark the spots of the pigments by circling with a pencil. Note the colors of the spots. Measure the distance of the center of each spot from its origin. Calculate the R_f values.

14. If the spots on the chromatogram are faded, we can visualize them by exposing the chromatogram to iodine vapor. Place your chromatogram into a wide-mouthed jar containing a few iodine crystals. Cap the jar and warm it slightly on a hot plate to enhance the sublimation of iodine. The iodine vapor will interact with the faded

pigment spots and make them visible. After a few minutes of exposure to iodine vapor, remove the chromatogram and mark the spots **immediately** with pencil. The spots will fade again with exposure to air. Measure the distance of the center of the spots from the origin and calculate the R_f values.

15. Record the results of the paper chromatography on the Report Sheet.

Chemicals and Equipment

1. Melting point capillaries open at both ends
2. 25-mL buret or chromatographic column
3. Glass wool
4. Whatman no.1 filter paper, 10 × 20 cm, cut to size
5. Heat lamp (optional)
6. Stapler
7. Hot plate (with or without water bath)
8. Tomato paste
9. Aluminum oxide (alumina)
10. Petroleum ether (b.p. 30–60°C)
11. 95% ethanol
12. Toluene
13. Acetone
14. 0.5% β-carotene in petroleum ether
15. Saturated bromine water
16. Iodine crystals
17. Ruler
18. Wide-mouthed jar

Identification of alcohols and phenols

Background

Specific groups of atoms in an organic molecule can determine its physical and chemical properties. These groups are referred to as *functional groups*. Organic compounds which contain the functional group $-OH$, the hydroxyl group, are called *alcohols*.

Alcohols are important commercially and include uses as solvents, drugs, and disinfectants. The most widely used alcohols are methanol or methyl alcohol, CH_3OH, ethanol or ethyl alcohol, CH_3CH_2OH, and 2-propanol or isopropyl alcohol, $(CH_3)_2CHOH$. Methyl alcohol is found in automotive products such as antifreeze and "dry gas." Ethyl alcohol is used as a solvent for drugs and chemicals, but is more popularly known for its effects as an alcoholic beverage. Isopropyl alcohol, also known as "rubbing alcohol," is an antiseptic.

Alcohols may be classified as either primary, secondary, or tertiary:

$$R-CH_2-OH$$
Primary alcohol

$$R-\underset{\underset{OH}{|}}{CH}-R'$$
Secondary alcohol

$$R-\underset{\underset{R''}{|}}{\overset{\overset{R'}{|}}{C}}-OH$$
Tertiary alcohol

Note that the classification depends on the number of carbon-containing groups, R (alkyl or aromatic), attached to the carbon bearing the hydroxyl group. Examples of each type are as follows:

$$CH_3CH_2-OH$$
Ethanol
(Ethyl alcohol)
a primary alcohol

$$CH_3-\underset{\underset{CH_3}{|}}{CH}-OH$$
2-Propanol
(Isopropyl alcohol)
a secondary alcohol

$$CH_3-\underset{\underset{CH_3}{|}}{\overset{\overset{CH_3}{|}}{C}}-OH$$
2-Methyl-2-propanol
(*t*-Butyl alcohol)
a tertiary alcohol

Phenols bear a close resemblance to alcohols structurally since the hydroxyl group is present. However, since the $-OH$ group is bonded directly to a carbon that is part of an aromatic ring, the chemistry is quite different from that of alcohols. Phenols are more acidic than alcohols; concentrated solutions of the compound phenol are quite toxic and can cause severe skin burns. Phenol derivatives are found in medicines; for example, thymol is used to kill fungi and hookworms. (Also see Table 5.1.)

Phenol

Thymol
(2-isopropyl-5-methylphenol)

In this experiment, you will examine physical and chemical properties of representative alcohols and phenols. You will be able to compare the differences in chemical behavior between these compounds and use this information to identify an unknown.

Table 5.1	Selected Alcohols and Phenols
Compound	**Name and Use**
CH_3OH	Methanol: solvent for paints, shellacs, and varnishes
CH_3CH_2OH	Ethanol: alcoholic beverages; solvent for medicines, perfumes, and varnishes
$CH_3 - \underset{\underset{OH}{\vert}}{CH} - CH_3$	Isopropyl alcohol (2-propanol): rubbing alcohol; astringent; solvent for cosmetics, perfumes, and skin creams
$\underset{\underset{OH}{\vert}}{CH_2} - \underset{\underset{OH}{\vert}}{CH_2}$	Ethylene glycol: antifreeze
$\underset{\underset{OH}{\vert}}{CH_2} - \underset{\underset{OH}{\vert}}{CH} - \underset{\underset{OH}{\vert}}{CH_2}$	Glycerol (glycerin): sweetening agent; solvent for medicines; lubricant; moistening agent
	Phenol (carbolic acid): cleans surgical and medical instruments; topical antipruritic (relieves itching)
	Vanillin: flavoring agent (vanilla flavor)
	Tetrahydrourushiol: irritant in poison ivy

Physical Properties

Since the hydroxyl group is present in alcohols and phenols, these compounds are polar. The polarity of the hydroxyl group, coupled with its ability to form hydrogen bonds, enables many alcohols and phenols to mix with water. Since these compounds also contain nonpolar portions, they show additional solubility in many organic solvents, such as dichloromethane and diethyl ether.

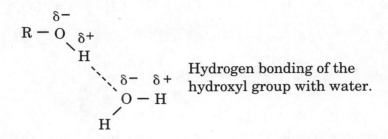

Hydrogen bonding of the hydroxyl group with water.

Chemical Properties

The chemical behavior of the different classes of alcohols and of phenols can be used as a means of identification. Quick, simple tests that can be carried out in test tubes will be performed.

1. *Lucas test.* This test is used to distinguish between water-soluble primary, secondary, and tertiary alcohols. Lucas reagent is a mixture of zinc chloride, $ZnCl_2$, in concentrated HCl. Upon addition of this reagent, a tertiary alcohol reacts rapidly and immediately gives an insoluble white layer. A secondary alcohol reacts slowly and, after heating slightly, gives the white layer within 10 min. A primary alcohol does not react. Any formation of a heterogeneous phase or appearance of an emulsion is a positive test.

$$CH_3CH_2—OH \ + \ HCl \ + \ ZnCl_2 \ \rightarrow \ \text{no reaction}$$
primary alcohol

$$(CH_3)_2CH—OH \ + \ HCl \ + \ ZnCl_2 \ \rightarrow \ (CH_3)_2CH—Cl \downarrow + H_2O \ \text{(10 min. heat)}$$
secondary alcohol   insoluble

$$(CH_3)_3C—OH \ + \ HCl \ + \ ZnCl_2 \ \rightarrow \ (CH_3)_3C—Cl \downarrow + H_2O \ \text{(<5 min.)}$$
tertiary alcohol   insoluble

2. *Chromic acid test.* This test is able to distinguish primary and secondary alcohols from tertiary alcohols. Using acidified dichromate solution, primary alcohols are oxidized to carboxylic acids; secondary alcohols are oxidized to ketones; tertiary alcohols are not oxidized. (Note that in those alcohols which are oxidized, the carbon that has the hydroxyl group *loses a hydrogen.*) In the oxidation, the brown-red color of the chromic acid changes to a blue-green solution. Phenols are oxidized to nondescript brown tarry masses. (Aldehydes are also oxidized under these conditions to carboxylic acids, but ketones remain intact; see Experiment 6 for further discussion.)

$$3CH_3CH_2-OH + 4H_2CrO_4 + 6H_2SO_4 \longrightarrow 3CH_3-\overset{\overset{\displaystyle O}{\|}}{C}-OH + 2Cr_2(SO_4)_3 + 13H_2O$$

primary alcohol   brown-red   carboxylic acid   blue-green

$$3CH_3-\overset{\overset{\displaystyle OH}{|}}{CH}-CH_3 + 2H_2CrO_4 + 3H_2SO_4 \longrightarrow 3CH_3-\overset{\overset{\displaystyle O}{\|}}{C}-CH_3 + Cr_2(SO_4)_3 + 8H_2O$$

secondary alcohol   brown-red   ketone   blue-green

$$(CH_3)_3C-OH + H_2CrO_4 + H_2SO_4 \longrightarrow \text{no reaction}$$

tertiary alcohol

3. *Iodoform test.* This test is more specific than the previous two tests. Only ethanol (ethyl alcohol) and alcohols with the part structure $CH_3CH(OH)$ react. These alcohols react with iodine in aqueous sodium hydroxide to give the yellow precipitate iodoform.

$$\overset{\overset{\displaystyle OH}{|}}{RCHCH_3} + 4I_2 + 6NaOH \longrightarrow \overset{\overset{\displaystyle O}{\|}}{RC}-O^-\,Na^+ + 5NaI + 5H_2O + HCI_3\,(s)$$

iodoform
yellow

Phenols also react under these conditions. With phenol, the yellow precipitate triiodophenol forms.

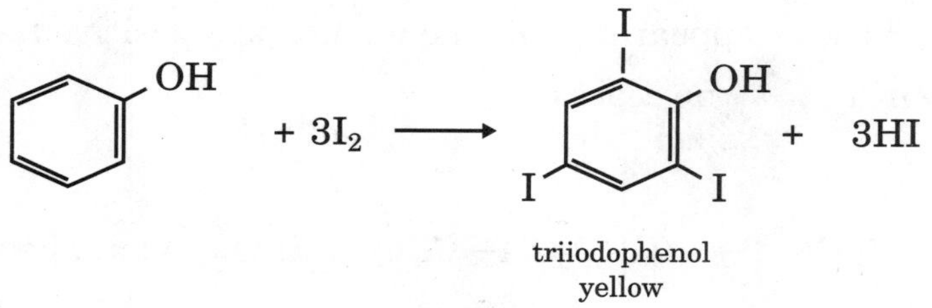

triiodophenol
yellow

4. *Acidity of phenol.* Phenol is also called carbolic acid. Phenol is an acid and will react with base; thus phenols readily dissolve in base solutions. In contrast, alcohols are not acidic.

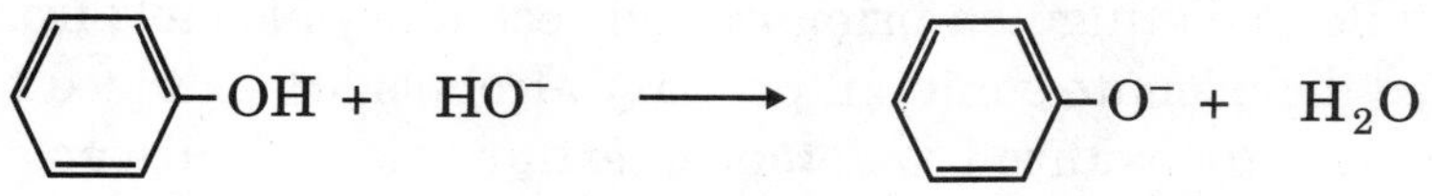

5. *Iron(III)chloride test.* Addition of aqueous iron(III) chloride to a phenol gives a colored solution. Depending on the structure of the phenol, the color can vary from green to purple.

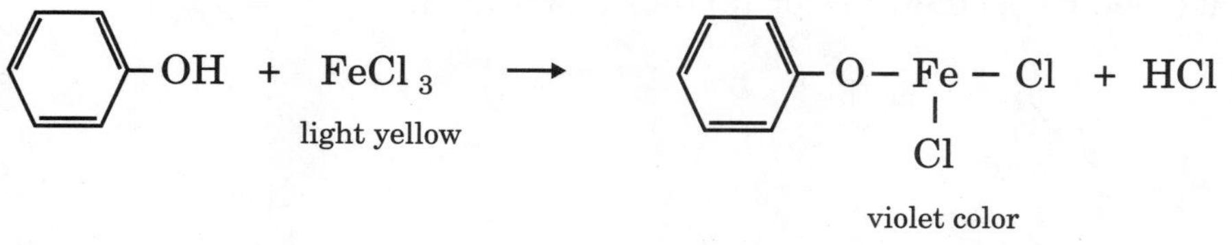

light yellow

violet color

1. To learn characteristic chemical reactions of alcohols and phenols.
2. To use these chemical characteristics for identification of an organic compound.

$1^° \quad 2^° \quad 3^°$

Procedure

CAUTION!

Chromic acid is very corrosive. Any spill should be immediately flushed with water. Phenol is toxic. Also, contact with the solid will cause burns to skin; any contact should be thoroughly washed with large quantities of water. Solid phenol should be handled only with a spatula or forceps. Use gloves with these reagents. Dispose of reaction mixtures and excess reagents in proper containers as directed by your instructor.

Physical Properties of Alcohols and Phenols

1. You will test the alcohols 1-butanol (a primary alcohol), 2-butanol (a secondary alcohol), 2-methyl-2-propanol (a tertiary alcohol), and phenol; you will also have as an unknown one of these compounds (labeled A, B, C, or D). As you run a test on a known, test the unknown at the same time for comparison. Note that the phenol will be provided as an aqueous solution.

2. Into separate test tubes (100 × 13 mm) labeled 1-butanol, 2-butanol, 2-methyl-2-propanol, *tert-butyl* and unknown, place 10 drops of each sample; dilute by mixing with 3 mL of distilled water. Into a separate test tube, place 2 mL of a prepared water solution of phenol. Are all the solutions homogeneous? Record your observations on the Report Sheet (1).

3. Test the pH of each of the aqueous solutions. Do the test by first dipping a clean glass rod into the solutions and then transferring a drop of liquid to pH paper. Use a broad pH indicator paper (e.g., pH range 1–12) and read the value of the pH by comparing the color to the chart on the dispenser. Record the results on the Report Sheet (2).

Chemical Properties of Alcohols and Phenols

1. *Iodoform test*. Place into separate clean, dry test tubes (150 × 18 mm), labeled 1-butanol, 2-butanol, 2-methyl-2-propanol, phenol, and unknown, 5 drops of sample to be tested. Add to each test tube 2 mL of water. If the compound is not soluble, add dioxane (dropwise) until the solution is homogeneous. Add to each test tube (dropwise) 2 mL of 6 M NaOH; tap the test tube with your finger to mix. The mixture is warmed in a 60°C water bath, and the prepared solution of I_2-KI test reagent is added dropwise (with shaking) until the solution becomes brown (approx. 25 drops). (If the color fades, add more I_2-KI test reagent until the dark color persists for 2 min. at 60°C.) Add 6 M

5 drops + H_2O (2 drops) + (dioxane) + 2 mL NaOH + I_2-KI

NaOH (dropwise) until the solution becomes colorless. Keep the test tubes in the warm water bath for 5 min. Remove the test tubes from the water, let cool, and look for a light yellow precipitate. Record your observations on the Report Sheet (3). The formation of the yellow precipitate tends to be slow. Put these test tubes to one side and make your observations when all the other tests are completed.

2. *Lucas test*. Place 5 drops of each sample into separate clean, dry test tubes (100 × 13 mm), labeled as before. Add 1 mL of Lucas reagent; mix well by stoppering each test tube with a cork, tapping the test tube sharply with your finger for a few seconds to mix; remove the cork after mixing and allow each test tube to stand for 5 min. Look carefully for any cloudiness that may develop during this time period. If there is no cloudiness after 10 min., warm the test tubes that are clear for 15 min. in a 60°C water bath. Record your observations on the Report Sheet (4).

3. *Chromic acid test*. Place into separate clean, dry test tubes (100 × 13 mm), labeled as before, 5 drops of sample to be tested. To each test tube add 10 drops of reagent grade acetone and 2 drops of chromic acid. Place the test tubes in a 60°C water bath for 5 min. Note the color of each solution. (Remember, the loss of the brown-red and the formation of a blue-green color is a positive test.) Record your observations on the Report Sheet (5).

4. *Iron(III) chloride test*. Place into separate clean, dry test tubes (100 × 13 mm), labeled as before, 5 drops of sample to be tested. Add 2 drops of iron(III) chloride solution to each. Note any color changes in each solution. (Remember, a purple color indicates the presence of a phenol.) Record your observations on the Report Sheet (6).

5. From your observations identify your unknown.

Chemicals and Equipment

1. Aqueous phenol
2. Acetone (reagent grade)
3. 1-Butanol
4. 2-Butanol
5. 2-Methyl-2-propanol (*t*-butyl alcohol)
6. Chromic acid solution
7. Dioxane
8. Iron(III) chloride solution
9. I_2-KI solution
10. Lucas reagent
11. Corks
12. Hot plate
13. pH paper
14. Unknown

Identification of aldehydes and ketones

Background

Aldehydes and ketones are representative of compounds which possess the carbonyl group:

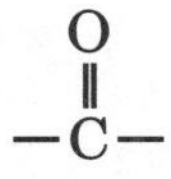

$$\overset{\displaystyle O}{\underset{\displaystyle |}{\overset{\displaystyle \|}{-C-}}}$$

The carbonyl group

Aldehydes have at least one hydrogen attached to the carbonyl carbon; in ketones, no hydrogens are directly attached to the carbonyl carbon, only carbon containing R-groups:

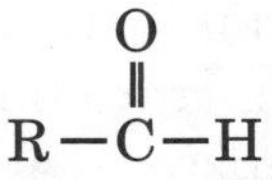

$$\overset{\displaystyle O}{\overset{\displaystyle \|}{R-C-H}}$$

Aldehyde

$$\overset{\displaystyle O}{\overset{\displaystyle \|}{R-C-R'}}$$

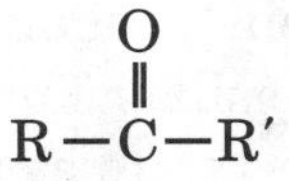

Ketone

(R and R' can be alkyl or aromatic)

Aldehydes and ketones of low molecular weight have commercial importance. Many others occur naturally. Table 6.1 has some representative examples.

Table 6.1	Representative Aldehydes and Ketones	
Compound		**Source and Use**
$\overset{O}{\overset{\|}{HCH}}$	Formaldehyde	Oxidation of methanol; plastics; preservative
$\overset{O}{\overset{\|}{CH_3CCH_3}}$	Acetone	Oxidation of isopropyl alcohol; solvent
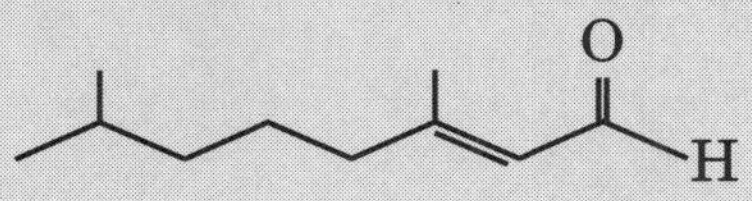	Citral	Lemongrass oil; fragrance
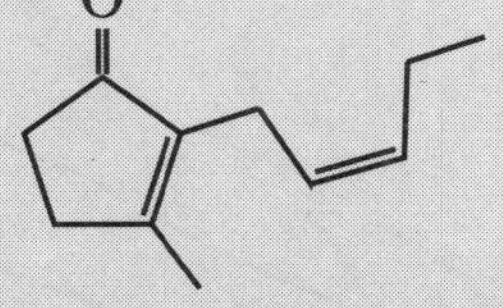	Jasmone	Oil of jasmine; fragrance

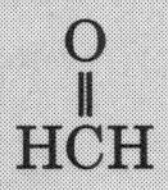

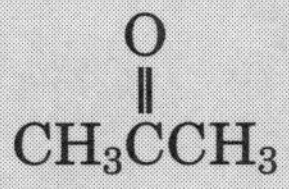

In this experiment you will investigate the chemical properties of representative aldehydes and ketones.

Classification Tests

1. *Chromic acid test*. Aldehydes are oxidized to carboxylic acids by chromic acid; ketones are not oxidized. A positive test results in the formation of a blue-green solution from the brown-red color of chromic acid.

$$3R-\overset{\overset{\displaystyle O}{\|}}{C}-H \ + \ 2H_2CrO_4 \ + \ 3H_2SO_4 \ \longrightarrow \ 3R-\overset{\overset{\displaystyle O}{\|}}{C}-OH \ + \ Cr_4(SO_4)_3 \ + \ 5H_2O$$

$$\underset{\text{aldehyde}}{} \qquad \underset{\text{brown-red}}{} \qquad\qquad\qquad\qquad \underset{\text{blue-green}}{}$$

$$\underset{\text{ketone}}{R-\overset{\overset{\displaystyle O}{\|}}{C}-R} \ \xrightarrow[\text{H}_2\text{SO}_4]{\text{H}_2\text{CrO}_4} \ \text{no reaction}$$

2. *Tollens' test*. Most aldehydes reduce Tollens' reagent (ammonia and silver nitrate) to give a precipitate of silver metal. The free silver forms a silver mirror on the sides of the test tube. (This test is sometimes referred to as the "silver mirror" test.) The aldehyde is oxidized to a carboxylic acid.

$$\underset{\text{aldehyde}}{R-\overset{\overset{\displaystyle O}{\|}}{C}-H} \ + \ 2Ag(NH_3)_2OH \ \longrightarrow \ \underset{\substack{\text{silver}\\\text{mirror}}}{2Ag(s)} \ + \ R-\overset{\overset{\displaystyle O}{\|}}{C}-O^-NH_4{}^+ \ + \ H_2O \ + \ 3NH_3$$

3. *Iodoform test*. Methyl ketones give the yellow precipitate iodoform when reacted with iodine in aqueous sodium hydroxide.

$$\underset{\substack{\text{methyl}\\\text{ketone}}}{R-\overset{\overset{\displaystyle O}{\|}}{C}-CH_3} \ + \ 3I_2 \ + \ 4NaOH \longrightarrow 3NaI \ + \ 3H_2O \ + \ R-\overset{\overset{\displaystyle O}{\|}}{C}-O^-Na^+ \ + \ \underset{\substack{\text{iodoform}\\\text{yellow}}}{HCI_3(s)}$$

4. *2,4-Dinitrophenylhydrazine test.* All aldehydes and ketones give an immediate precipitate with 2,4-dinitrophenylhydrazine reagent. This reaction is general for both these functional groups. The color of the precipitate varies from yellow to red. (Note that alcohols do not give this test.)

Identification by forming a derivative

The classification tests (summarized in Table 6.2), when properly done, can distinguish between various types of aldehydes and ketones. However, these tests alone may not allow for the identification of a specific unknown aldehyde or ketone. A way to correctly identify an unknown compound is by using a known chemical reaction to convert it into another compound that is known. The new compound is referred to as a *derivative*. Then, by comparing the physical properties of the unknown and the derivative to the physical properties of known compounds listed in a table, an identification can be made.

Table **6.2** Summary of Classification Tests	
Compound	**Reagent for Positive Test**
Aldehydes and ketones	2,4-Dinitrophenylhydrazine
Aldehydes	Chromic acid
	Tollens' reagent
Methyl ketones	Iodoform

The ideal derivative is a solid. A solid can be easily purified by crystallization and easily characterized by its melting point. Thus two similar aldehydes or two similar ketones usually have derivatives that have *different melting points*. The most frequently formed derivatives for aldehydes and ketones are the 2,4-dinitrophenylhydrazone

(2,4-DNP), oxime, and semicarbazone. Table 6.4 (p. 76) lists some aldehydes and ketones along with melting points of their derivatives. If, for example, we look at the properties of valeraldehyde and crotonaldehyde, though the boiling points are virtually the same, the melting points of the 2,4-DNP, oxime, and semicarbazone are different and provide a basis for identification.

1. *2,4-Dinitrophenylhydrazone*. 2,4-Dinitrophenylhydrazine reacts with aldehydes and ketones to form 2,4-dinitrophenylhydrazones (2,4-DNP).

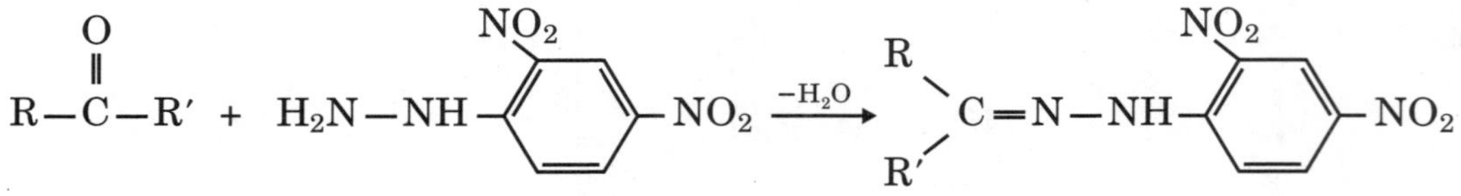

2,4-dinitrophenylhydrazine   2,4-dinitrophenylhydrazone (2,4-DNP)

(R′ = H or alkyl or aromatic)

The 2,4-DNP product is usually a colored solid (yellow to red) and is easily purified by recrystallization.

2. *Oxime*. Hydroxylamine reacts with aldehydes and ketones to form oximes.

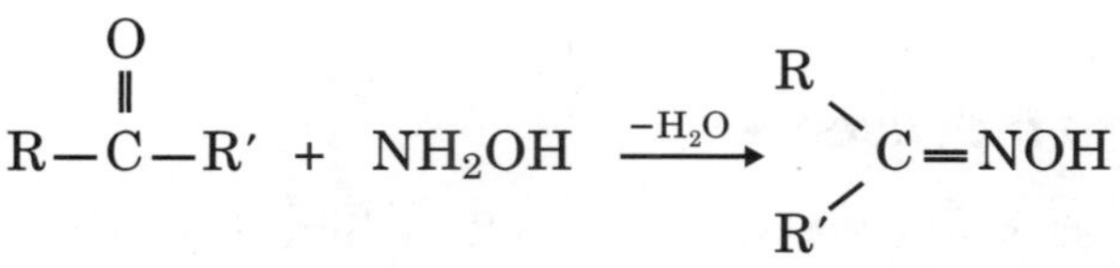

hydroxylamine   oxime

(R′ = H or alkyl or aromatic)

These are usually derivatives with low melting points.

3. *Semicarbazone*. Semicarbazide, as its hydrochloride salt, reacts with aldehydes and ketones to form semicarbazones.

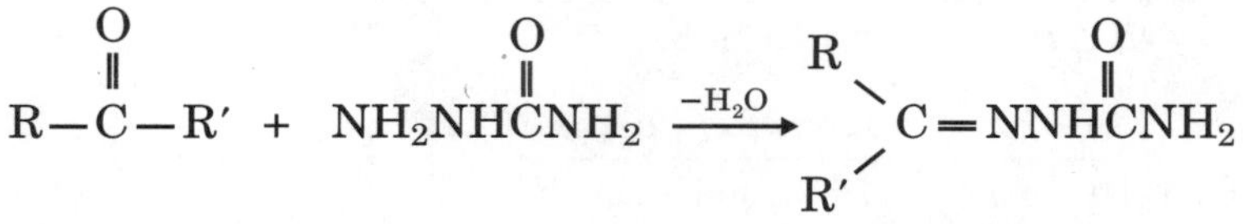

semicarbazide   semicarbazone

(R′ = H or alkyl or aromatic)

A pyridine base is used to neutralize the hydrochloride in order to free the semicarbazide so it may react with the carbonyl substrate.

Objectives

1. To learn the chemical characteristics of aldehydes and ketones.
2. To use these chemical characteristics in simple tests to distinguish between examples of aldehydes and ketones.
3. To identify aldehydes and ketones by formation of derivatives.

Classification Tests

1. Classification tests are to be carried out on four known compounds and one unknown. Any one test should be carried out on all five samples at the same time for comparison. Label test tubes as shown in Table 6.3.

Table 6.3	Labelling Test Tubes
Test Tube No.	**Compound**
1	Isovaleraldehyde (an aliphatic aldehyde)
2	Benzaldehyde (an aromatic aldehyde)
3	Cyclohexanone (a ketone)
4	Acetone (a methyl ketone)
5	Unknown

CAUTION!

Chromic acid is toxic and corrosive. Handle with care and **promptly wash any spill**. Use gloves with this reagent.

2. *Chromic acid test.* Place 5 drops of each substance into separate, labeled test tubes (100×13 mm). Dissolve each compound in 20 drops of reagent-grade acetone (to serve as solvent); then add to each test tube 4 drops of chromic acid reagent, one drop at a time; after each drop, mix by sharply tapping the test tube with your finger. Let stand for 10 min. Aliphatic aldehydes should show a change within a minute; aromatic aldehydes take longer. Note the approximate time for any change in color or formation of a precipitate on the Report Sheet.

3. *Tollens' test.*

CAUTION!

The reagent must be freshly prepared before it is to be used and any excess disposed of **immediately** after use. Organic residues should be discarded in appropriate waste containers. Unused Tollens' reagent should be collected from every student by the instructor. **Do not store Tollens' reagent since it is explosive when dry.** The instructor should dispose of the excess reagant by adding 1 M HNO_3 until acidic, warming on a hot plate. The solution can then be stored in a waste container for heavy metals.

Enough reagent for your use can be prepared in a 25-mL Erlenmeyer flask by mixing 5 mL of Tollens' solution A with 5 mL of Tollens' solution B. To the silver oxide precipitate which forms, add (dropwise, with shaking) 10% ammonia solution until the brown precipitate just dissolves. *Avoid an excess of ammonia.*

Place 5 drops of each substance into separately labeled clean, dry test tubes (100 × 13 mm). ~~Dissolve the compound in~~ *bis*~~(2-ethoxyethyl)ether by adding this~~ ~~solvent dropwise until a homogeneous solution is obtained.~~ Then, add 2 mL (approx. 40 drops) of the prepared Tollens' reagent and mix by sharply tapping the test tube with your finger. Place the test tube in a 60°C water bath for 5 min. Remove the test tubes from the water and look for a silver mirror. If the tube is clean, a silver mirror will be formed; if not, a black precipitate of finely divided silver will appear. Record your results on the Report Sheet. (Clean your test tubes with 1 M HNO_3 and discard the solution in a waste container designated by your instructor.)

4. *Iodoform test*. Place 5 drops of each sample into separate clean, dry test tubes (150 × 18 mm). Add to each test tube 2mL of water. If the compound is not soluble, add dioxane (dropwise) until the solution is homogeneous. Add to each test tube (dropwise) 2mL of 6 M NaOH; tap the test tube with your finger to mix. The mixture is warmed in a 60°C water bath, and the prepared solution of I_2-KI test reagent is added dropwise (with shaking) until the solution becomes brown (approximately 25 drops). (If the color fades, add more I_2-KI test reagent until the dark color persists for 2 min. at 60°C.) Add 6 M NaOH (dropwise) until the solution becomes colorless. Keep the test tubes in the warm water bath for 5 min. Remove the test tubes from the water, let cool, and look for a yellow precipitate. Record your observations on the Report Sheet. The formation of the yellow precipitate tends to be slow. Put these test tubes to one side and make your observations when all the other tests are completed.

5. *2,4-Dinitrophenylhydrazine test*. Place 5 drops of each substance into separately labeled clean, dry test tubes (100 × 13 mm) and add 20 drops of the 2,4-dinitrophenylhydrazine reagent to each. If no precipitate forms immediately, heat for 5 min. in a warm water bath (60°C); cool. Record your observations on the Report Sheet.

Formation of derivatives

CAUTION!

The chemicals used to prepare derivatives and some of the derivatives are potential carcinogens. Exercise care in using the reagents and in handling the derivatives. Avoid skin contact by wearing gloves.

---

1. This section is *optional*. Consult your instructor to determine whether this section is to be completed. Your instructor will indicate how many derivatives and which derivatives you should make.

2. *Waste*. Place all the waste solutions from these preparations in designated waste containers for disposal by your instructor.

3. *General procedure for recrystallization*. Heat a small volume (10–20 mL) of solvent to boiling on a steam bath (or carefully on a hot plate). Place crystals into a test tube (100 × 13 mm) and add the hot solvent (dropwise) until the crystals just dissolve (keep the solution hot, also). Allow the solution to cool to room temperature; then cool further in an ice bath. Collect the crystals on a Hirsch funnel by vacuum filtration. Use a trap between the Hirsch funnel set-up and the aspirator (Fig. 6.1). Wash the

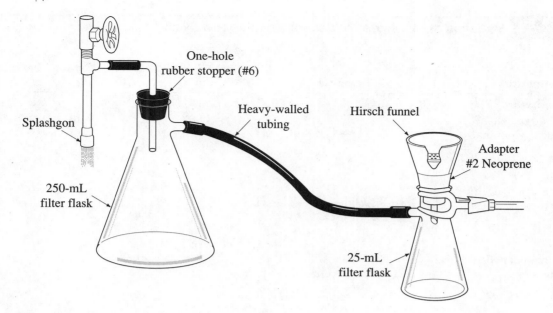

Figure 6.1 • Vacuum filtration with a Hirsch funnel.

crystals with 10 drops of ice cold solvent. Allow the crystals to dry by drawing air through the Hirsch funnel. Take a melting point.

4. *2,4-Dinitrophenylhydrazone* (2,4-DNP). Place 5 mL of the 2,4-dinitrophenylhydrazine reagent in a test tube (150 × 18 mm). Add 10 drops of the unknown compound; sharply tap the test tube with your finger to mix. If crystals do not form immediately, gently heat in a water bath (60°C) for 5 min. Cool in an ice bath until crystals form. Collect the crystals by vacuum filtration using a Hirsch funnel (Fig. 6.1). Allow the crystals to dry on the Hirsch funnel by drawing air through the crystals. Take a melting point and record on the Report Sheet. (The crystals are usually pure enough to give a good melting point. However, if the melting point range is too large, recrystallize from a minimum volume of ethanol.)

5. *Oxime*. Prepare fresh reagent by dissolving 1.0 g of hydroxylamine hydrochloride and 1.5 g of sodium acetate in 4 mL of distilled water in a test tube (150 × 18 mm). Add 20 drops of unknown and sharply tap the test tube with your finger to mix. Warm in a hot water bath (60°C) for 5 min. Cool in an ice bath until crystals form. (If no crystals form, scratch the inside of the test tube with a glass rod.) Collect the crystals on a Hirsch funnel by vacuum filtration (Fig. 6.1). Allow the crystals to air dry on the Hirsch funnel by drawing air through the crystals. Take a melting point and record on the Report Sheet. (Recrystallize, if necessary, from a minimum volume of ethanol.)

6. *Semicarbazone*. Place 2.0 mL of the semicarbazide reagent in a test tube (150 × 18 mm); add 10 drops of unknown. If the solution is not clear, add methanol (dropwise) until a clear solution results. Add 2.0 mL of pyridine and gently warm in a hot bath (60°C) for 5 min. Crystals should begin to form. (If there are no crystals, place the test tube in an ice bath and scratch the inside of the test tube with a glass rod.) Collect the crystals on a Hirsch funnel by vacuum filtration (Fig. 6.1). Allow the crystals to air dry on the Hirsch funnel by drawing air through the crystals. Take a melting point and record on the Report Sheet. (Recrystallize, if necessary, from a minimum volume of ethanol.)

7. Based on the observations you recorded on the Report Sheet, and by comparing the melting points of the derivatives for your unknown to the knowns listed in Table 6.4, identify your unknown.

Table **6.4** **Selection of Aldehydes and Ketones with Derivatives**

Compound	Formula	b.p.°C	2,4-DNP m.p.°C	Oxime m.p.°C	Semi-carbazone m.p.°C
Aldehydes					
Isovaleraldehyde (3-methylbutanal)	CH_3 $\mid$ $CH_3-CH-CH_2-C-H$ ($\parallel$ O)	93	123	49	107
Valeraldehyde (pentanal)	$CH_3CH_2CH_2CH_2-C-H$ ($\parallel$ O)	103	106	52	—
Crotonaldehyde (2-butenal)	$CH_3-CH=CH-C-H$ ($\parallel$ O)	104	190	119	199
Caprylaldehyde (octanal)	$CH_3CH_2CH_2CH_2CH_2CH_2CH_2-C-H$ ($\parallel$ O)	171	106	60	101
Benzaldehyde	C_6H_5-C-H ($\parallel$ O)	178	237	35	222
Ketones					
Acetone (2-propanone)	CH_3-C-CH_3 ($\parallel$ O)	56	126	59	187
2-Pentanone	$CH_3-C-CH_2CH_2CH_3$ ($\parallel$ O)	102	144	58 (b.p. 167°C)	112
3-Pentanone	$CH_3CH_2-C-CH_2CH_3$ ($\parallel$ O)	102	156	69 (b.p. 165°C)	139
Cyclopentanone	(cyclopentanone structure)	131	146	56	210
Cyclohexanone	(cyclohexanone structure)	156	162	90	166
Acetophenone	$C_6H_5-C-CH_3$ ($\parallel$ O)	202	238	60	198

Source: Compiled by Zvi Rappoport, *CRC Handbook of Tables for Organic Compound Identification, 3rd ed.*, The Chemical Rubber Co.: Cleveland (1967).

Chemicals and Equipment

1. Acetone (reagent grade)
2. 10% ammonia solution
3. Benzaldehyde
4. *Bis*(2-ethoxyethyl) ether
5. Chromic acid reagent
6. Cyclohexane
7. 2,4-Dinitrophenylhydrazine reagent
8. Dioxane
9. Ethanol
10. Hydroxylamine hydrochloride
11. I₂-KI test solution
12. Isovaleraldehyde
13. Methanol
14. 6 M NaOH, sodium hydroxide
15. Pyridine
16. Semicarbazide reagent
17. Sodium acetate
18. Tollens' reagent (solution A and solution B)
19. Hirsch funnel
20. Hot plate
21. Neoprene adapter (no. 2)
22. Rubber stopper (no. 6, one-hole), with glass tubing
23. 50-mL side-arm filter flask
24. 250-mL side-arm filter flask
25. Vacuum tubing (heavy walled)

Chromic Test

+ H_2CrO_4 + H_2SO_4
brown-
red
(not organic,
Chromium changes
color)

$\rightarrow Cr_2(SO_4)_3 + 5H_2O$
blue
-green

$+ CH_3-CH-CH_2-C-OH$
(CH₃ on carbon, O on carbonyl)
oxidized

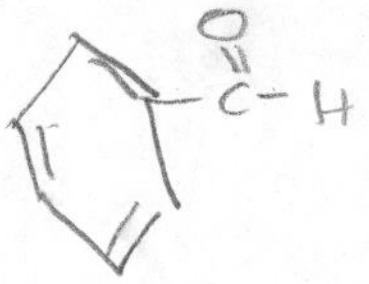

$\rightarrow$

Exploring
how ald. + ketones
react.
methyl ketones react —
Aldehydes react

Aldehyde

$\rightarrow$

+ 2 Ag(NH₃)₂OH

$\rightarrow$ 2Ag + H_2O + NH₃
Silver   ammonia
+   (poison)

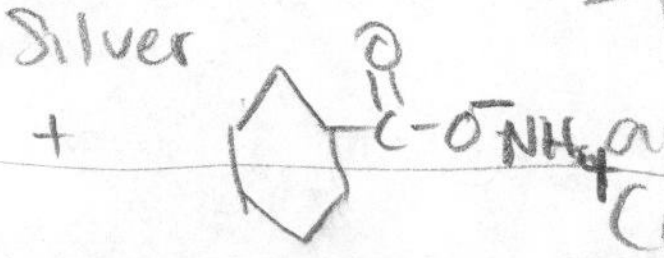

Properties of carboxylic acids and esters

Background

Carboxylic acids are structurally like aldehydes and ketones in that they contain the carbonyl group. However, an important difference is that carboxylic acids contain a hydroxyl group attached to the carbonyl carbon.

$$\overset{\displaystyle O}{\overset{\|}{-C}}-OH \qquad \text{The carboxylic acid group}$$

This combination gives the group its most important characteristic; it behaves as an acid.

   As a family, carboxylic acids are weak acids that ionize only slightly in water. As aqueous solutions, typical carboxylic acids ionize to the extent of only one percent or less.

$$\overset{\displaystyle O}{\overset{\|}{R-C}}-OH + H_2O \rightleftharpoons \overset{\displaystyle O}{\overset{\|}{R-C}}-O^- + H_3O^+$$

At equilibrium, most of the acid is present as un-ionized molecules. Dissociation constants, K_a, of carboxylic acids, where R is an alkyl group, are 10^{-5} or less. Water solubility depends to a large extent on the size of the R-group. Only a few low-molecular-weight acids (up to four carbons) are very soluble in water.

   Although carboxylic acids are weak, they are capable of reacting with bases stronger than water. Thus while benzoic acid shows limited water solubility, it reacts with sodium hydroxide to form the soluble salt sodium benzoate. (Sodium benzoate is a preservative in soft drinks.)

Benzoic acid   Sodium benzoate
Insoluble   Soluble

Sodium carbonate, Na_2CO_3, and sodium bicarbonate, $NaHCO_3$, solutions can neutralize carboxylic acids also.

The combination of a carboxylic acid and an alcohol gives an ester; water is eliminated. Ester formation is an equilibrium process, catalyzed by an acid catalyst.

$$CH_3CH_2CH_2\overset{\displaystyle O}{\overset{\|}{C}}-OH \;+\; CH_3CH_2OH \;\underset{}{\overset{H^+}{\rightleftharpoons}}\; H_2O \;+\; CH_3CH_2CH_2\overset{\displaystyle O}{\overset{\|}{C}}-OCH_2CH_3$$

Butyric acid   Ethyl alcohol   Ethyl butyrate (Ester)

Esterification $\longrightarrow$

$\longleftarrow$ Hydrolysis

The reaction typically gives 60% to 70% of the maximum yield. The reaction is a reversible process. An ester reacting with water, giving the carboxylic acid and alcohol, is called *hydrolysis*; it is acid catalyzed. The base-promoted decomposition of esters yields an alcohol and a salt of the carboxylic acid; this process is called *saponification*. Saponification means "soap making," and the sodium salt of a fatty acid (e.g., sodium stearate) is a soap.

$$CH_3CH_2CH_2\overset{\displaystyle O}{\overset{\|}{C}}-OCH_2CH_3 \;+\; NaOH \;\longrightarrow\; CH_3CH_2CH_2\overset{\displaystyle O}{\overset{\|}{C}}-O^-Na^+ \;+\; CH_3CH_2OH$$

Saponification

A distinctive difference between carboxylic acids and esters is in their characteristic odors. Carboxylic acids are noted for their sour, disagreeable odors. On the other hand, esters have sweet and pleasant odors often associated with fruits, and fruits smell the way they do because they contain esters. These compounds are used in the food industry as fragrances and flavoring agents. For example, the putrid odor of rancid butter is due to the presence of butyric acid, while the odor of pineapple is due to the presence of the ester, ethyl butyrate. Only those carboxylic acids of low molecular weight have odor at room temperature. Higher-molecular-weight carboxylic acids form strong hydrogen bonds, are solid, and have a low vapor pressure. Thus few molecules reach our noses. Esters, however, do not form hydrogen bonds among themselves; they are liquid at room temperature, even when the molecular weight is high. Thus they have high vapor pressure and many molecules can reach our noses, providing odor.

Objectives

1. To study the physical and chemical properties of carboxylic acids: solubility, acidity, aroma.
2. To prepare a variety of esters and note their odors.
3. To demonstrate saponification.

Procedure

Carboxylic Acids and Their Salts

Characteristics of acetic acid

1. Place into a clean, dry test tube (100 × 13 mm) 2 mL of water and 10 drops of glacial acetic acid. Note its odor by wafting (moving your hand quickly over the open end of the test tube) the vapors toward your nose. Of what does it remind you?

2. Take a glass rod and dip it into the solution. Using wide-range indicator paper (pH 1–12), test the pH of the solution by touching the paper with the wet glass rod. Determine the value of the pH by comparing the color of the paper with the chart on the dispenser.

3. Now, add 2 mL of 2 M NaOH to the solution. Cork the test tube and sharply tap it with your finger. Remove the cork and determine the pH of the solution as before; if not basic, continue to add more base (dropwise) until the solution is basic. Note the odor and compare to the odor of the solution before the addition of base.

4. By dropwise addition of 3 M HCl, carefully reacidify the solution from step no. 3 (above); test the solution as before with pH paper until the solution tests acid. Does the original odor return?

Characteristics of benzoic acid

1. Your instructor will weigh out 0.1 g of benzoic acid for sample size comparison. With your microspatula, take some sample equivalent to the preweighed sample (an exact quantity is not important here). Add the solid to a test tube (100 × 13 mm) along with 2 mL of water. Is there any odor? Mix the solution by sharply tapping the test tube with your finger. How soluble is the benzoic acid?

2. Now add 1 mL of 2 M NaOH to the solution from step no. 1 (above), cork, and mix by sharply tapping the test tube with your finger. What happens to the solid benzoic acid? Is there any odor?

3. By dropwise addition of 3 M HCl, carefully reacidify the solution from step no. 2 (above); test as before with pH paper until acidic. As the solution becomes acidic, what do you observe?

Esterification

1. Into five clean, dry test tubes (100 × 13 mm), add 10 drops of liquid carboxylic acid or 0.1 g of solid carboxylic acid and 10 drops of alcohol according to the scheme in Table 7.1. Note the odor of each reactant.

Table 7.1	Acids and Alcohols	
Test Tube No.	**Carboxylic Acid**	**Alcohol**
1	Formic	Isobutyl
2	Acetic	Benzyl
3	Acetic	Isopentyl
4	Acetic	Ethyl
5	Salicylic	Methyl

2. Add 5 drops of concentrated sulfuric acid to each test tube and mix the contents thoroughly by sharply tapping the test tube with your finger.

CAUTION!

Sulfuric acid causes severe burns. Flush any spill with lots of water. Use gloves with this reagent.

3. Place the test tubes in a warm water bath at 60°C for 15 min. Remove the test tubes from the water bath, cool, and add 2 mL of water to each. Note that there is a layer on top of the water in each test tube. With a Pasteur pipet, take a few drops from this top layer and place on a watch glass. Note the odor. Match the ester from each test tube with one of the following odors: banana, peach, raspberry, nail polish remover, wintergreen.

Saponification

This part of the experiment can be done while the esterification reactions are being heated.

1. Place into a test tube (150 × 18 mm) 10 drops of methyl salicylate and 5 mL of 6 M NaOH. Heat the contents in a boiling water bath for 30 min. Record on the Report Sheet what has happened to the ester layer (1).

2. Cool the test tube to room temperature by placing it in an ice water bath. Determine the odor of the solution and record your observation on the Report Sheet (2).

3. Carefully add 6 M HCl to the solution, 1 mL at a time, until the solution is acidic. After each addition, mix the contents and test the solution with litmus. When the solution is acidic, what do you observe? What is the name of the compound formed? Answer these questions on the Report Sheet (3).

Chemicals and Equipment

1. Glacial acetic acid
2. Benzoic acid
3. Formic acid
4. Salicylic acid
5. Benzyl alcohol
6. Ethanol (ethyl alcohol)
7. 2-Methyl-1-propanol (isobutyl alcohol)
8. 3-Methyl-1-butanol (isopentyl alcohol)
9. Methanol (methyl alcohol)
10. Methyl salicylate
11. 3 M HCl
12. 6 M HCl
13. 2 M NaOH
14. 6 M NaOH
15. Concentrated H_2SO_4
16. pH paper (broad range pH 1–12)
17. Litmus paper
18. Pasteur pipet
19. Hot plate

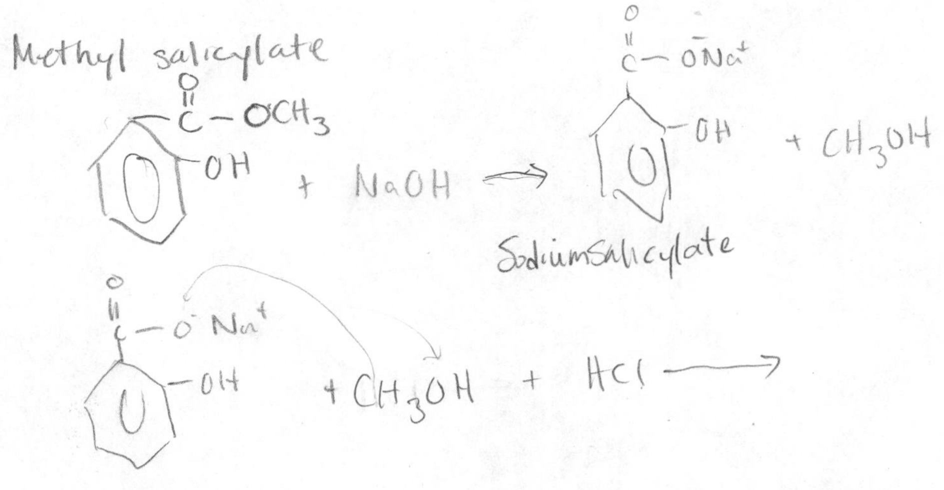

Methyl salicylate

Properties of amines and amides

Background

Amines and amides are two classes of organic compounds which contain nitrogen. Amines behave as organic bases and may be considered as derivatives of ammonia. Amides are compounds which have a carbonyl group connected to a nitrogen atom and are neutral. In this experiment, you will learn about the physical and chemical properties of some members of the amine and amide families.

If the hydrogens of ammonia are replaced by alkyl or aryl groups, amines result. Depending on the number of carbon atoms bonded directly to nitrogen, amines are classified as either primary (one carbon atom), secondary (two carbon atoms), or tertiary (three carbon atoms) (Table 8.1).

Table 8.1 Types of Amines

	Primary Amines	**Secondary Amines**	**Tertiary Amines**
NH_3 Ammonia	CH_3NH_2 Methylamine	$(CH_3)_2NH$ Dimethylamine	$(CH_3)_3N$ Trimethylamine
	Aniline	N-Methylaniline	N,N-Dimethylaniline

There are a number of similarities between ammonia and amines that carry beyond the structure. Consider odor. The smell of amines resembles that of ammonia but is not as sharp. However, amines can be quite pungent. Anyone handling or working with raw fish knows how strong the amine odor can be, since raw fish contains low-molecular-weight amines such as dimethylamine and trimethylamine. Other amines associated with decaying flesh have names suggestive of their odors: putrescine and cadaverine.

$$NH_2CH_2CH_2CH_2CH_2NH_2$$

Putrescine
(1,4-Diaminobutane)

$$NH_2CH_2CH_2CH_2CH_2CH_2NH_2$$

Cadaverine
(1,5-Diaminopentane)

The solubility of low-molecular-weight amines in water is high. In general, if the total number of carbons attached to nitrogen is four or less, the amine is water soluble; amines with a carbon content greater than four are water insoluble. However, all amines are soluble in organic solvents such as diethyl ether or dichloromethane.

Since amines are organic bases, water solutions show weakly basic properties. If the basicity of aliphatic amines and aromatic amines are compared to ammonia, aliphatic amines are stronger than ammonia, while aromatic amines are weaker. Amines characteristically react with acids to form ammonium salts; the nonbonded electron pair on nitrogen bonds the hydrogen ion.

$$RNH_2 + HCl \longrightarrow RNH_3^+Cl^-$$

Amine   Ammonium Salt

If an amine is insoluble, reaction with an acid produces a water-soluble salt. Since ammonium salts are water soluble, many drugs containing amines are prepared as ammonium salts. After working with fish in the kitchen, a convenient way to rid one's hands of fish odor is to rub a freshly cut lemon over the hands. The citric acid found in the lemon reacts with the amines found on the fish; a salt forms which can be easily rinsed away with water.

Amides are carboxylic acid derivatives. The amide group is recognized by the nitrogen connected to the carbonyl group. Amides are neutral compounds; the electrons are delocalized into the carbonyl (resonance) and thus, are not available to bond to a hydrogen ion.

Amide group   Acetamide   Benzamide

Under suitable conditions, amide formation can take place between an amine and a carboxylic acid, an acyl halide, or an acid anhydride. Along with ammonia, primary and secondary amines yield amides with carboxylic acids or derivatives. Table 8.2 relates the nitrogen base with the amide class (based on the number of alkyl or aryl groups on the nitrogen of the amide).

Table 8.2 Classes of Amides

Nitrogen Base	(reacts to form)	Amide ($-\overset{O}{\overset{\|}{C}}-\overset{\|}{N}-$)
Ammonia		Primary amide (no R groups)
Primary amine		Secondary amide (one R group)
Secondary amine		Tertiary amide (two R groups)

Hydrolysis of amides can take place in either acid or base. Primary amides hydrolyze in acid to ammonium salts and carboxylic acids. Neutralization of the acid and ammonium salts releases ammonia which can be detected by odor or by litmus.

$$R-\underset{\underset{\displaystyle O}{\|}}{C}-NH_2 + HCl + H_2O \longrightarrow R-\underset{\underset{\displaystyle O}{\|}}{C}-OH + NH_4Cl$$

$$NH_4Cl + NaOH \longrightarrow NH_3 + NaCl + H_2O$$

Secondary and tertiary amides would release the corresponding alkyl ammonium salts which, when neutralized, would yield the amine.

In base, primary amides hydrolyze to carboxylic acid salts and ammonia. The presence of ammonia (or amine from corresponding amides) can be detected similarly by odor or litmus. The carboxylic acid would be generated by neutralization with acid.

$$R-\underset{\underset{\displaystyle O}{\|}}{C}-NH_2 + NaOH \longrightarrow R-\underset{\underset{\displaystyle O}{\|}}{C}-O^-Na^+ + NH_3$$

$$R-\underset{\underset{\displaystyle O}{\|}}{C}-O^-Na^+ + HCl \longrightarrow R-\underset{\underset{\displaystyle O}{\|}}{C}-OH + NaCl$$

Objectives

1. To show some physical and chemical properties of amines and amides.
2. To demonstrate the hydrolysis of amides.

Procedure

CAUTION!

Amines are toxic chemicals. Avoid excessive inhaling of the vapors and use gloves to avoid direct skin contact. Anilines are more toxic than aliphatic amines and are readily absorbed through the skin. Wash any amine or aniline spill with large quantities of water. Diethyl ether (ether) is extremely flammable. Be certain there are **NO** open flames in the immediate area.

Properties of Amines

1. Place 5 drops of liquid or 0.1 g of solid from the compounds listed in the following table into labeled clean, dry test tubes (100 × 13 mm).

Test Tube No.	Nitrogen Compound
1	6 M NH$_3$
2	Triethylamine
3	Aniline
4	N,N-Dimethylaniline
5	Acetamide

2. Carefully note the odors of each compound. **Do not inhale deeply. Merely wave your hand across the mouth of the test tube toward your nose in order to note the odor.** Record your observations on the Report Sheet.

3. Add 2 mL of distilled water to each of the labeled test tubes. Mix thoroughly by sharply tapping the test tube with your finger. Note on the Report Sheet whether the amines are soluble or insoluble.

4. Take a glass rod, and test each solution for its pH. Carefully dip one end of the glass rod into a solution and touch a piece of pH paper. Between each test, be sure to clean and dry the glass rod. Record the pH by comparing the color of the paper with the chart on the dispenser.

5. Carefully add 2 mL of 6 M HCl to each test tube. Mix thoroughly by sharply tapping the test tube with your finger. Compare the odor and solubility of this solution to previous observations.

6. Place 5 drops of liquid or 0.1 g of solid from the compounds listed in the table into labeled clean, dry test tubes (100 × 13 mm). Add 2 mL of diethyl ether (ether) to each test tube. Stopper with a cork and mix thoroughly by shaking. Record the observed solubilities.

7. Carefully place on a watch glass, side-by-side, without touching, a drop of triethylamine and a drop of concentrated HCl. Record your observations.

Hydrolysis of Acetamide

1. Dissolve 0.5 g of acetamide in 5 mL of 6 M H$_2$SO$_4$ in a large test tube (150 × 18 mm). Heat the solution in a boiling water bath for 5 min.

2. Hold a small strip of moist pH paper over the mouth of the test tube; note any changes in color; record the pH reading. Remove the test tube from the water bath, holding it in a test tube holder. Carefully note any odor.

3. Place the test tube in an ice water bath until cool to the touch. Now *carefully add, dropwise with shaking*, 6 M NaOH to the cool solution until basic. (You will need more than 7 mL of base.) Hold a piece of moist pH paper over the mouth. Record the pH reading. Carefully note any odor.

Chemicals and Equipment

1. Acetamide
2. 6 M NH_3, ammonia water
3. Aniline
4. N,N-Dimethylaniline
5. Triethylamine
6. Diethyl ether (ether)
7. 6 M NaOH
8. Concentrated HCl
9. 6 M HCl
10. 6 M H_2SO_4
11. pH papers
12. Hot plate

Experiment 9

Polymerization reactions

Background

Polymers are giant molecules made of many (poly-) small units. The starting material, which is a single unit, is called the monomer. Many of the most important biological compounds are polymers. Cellulose and starch are polymers of glucose units, proteins are made of amino acids, and nucleic acids are polymers of nucleotides. Since the 1930s, a large number of synthetic polymers have been manufactured. They contribute to our comfort and gave rise to the previous slogan of DuPont Co.: "Better living through chemistry." Synthetic fibers such as nylon and polyesters, plastics such as the packaging materials made of polyethylene and polypropylene films, polystyrene, and polyvinyl chloride, just to name a few, all became household words. Synthetic polymers are parts of buildings, automobiles, machinery, toys, appliances, etc.; we encounter them daily in our life.

We focus our attention in this experiment on synthetic polymers and the basic mechanism by which some of them are formed. The two most important types of reactions that are employed in polymer manufacturing are the addition and condensation polymerization reactions. The first is represented by the polymerization of styrene and the second by the formation of nylon.

Styrene is a simple organic monomer which, by its virtue of containing a double bond, can undergo addition polymerization.

$$H_2C = CH + H_2C = CH \longrightarrow H_3C - CH - CH = CH$$

The reaction is called an *addition reaction* because two monomers are added to each other with the elimination of a double bond. This is also called a chain growth polymerization reaction. However, the reaction as such does not go without the help of an unstable molecule, called an *initiator*, that starts the reaction. Benzoyl peroxide or *t*-butyl benzoyl peroxide are such initiators. Benzoyl peroxide splits into two halves under the influence of heat or ultraviolet light and thus produces two free radicals. A *free radical* is a molecular fragment that has one unpaired electron. Thus, when the central bond was broken in the benzoyl peroxide, each of the shared pair of electrons went with one half of the molecule, each containing an unpaired electron.

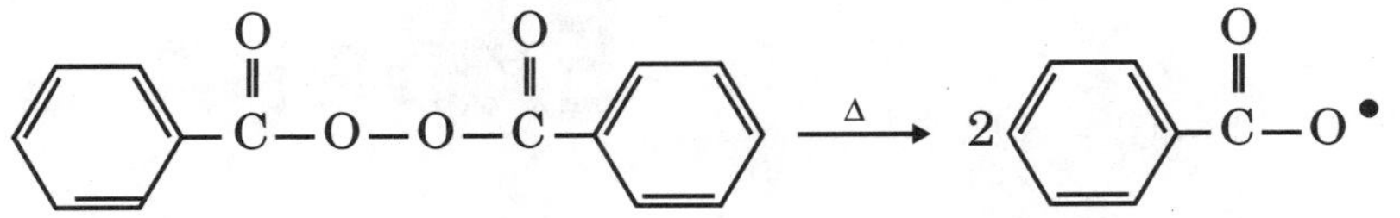

benzoyl peroxide

Similarly, *t*-butyl benzoyl peroxide also gives two free radicals:

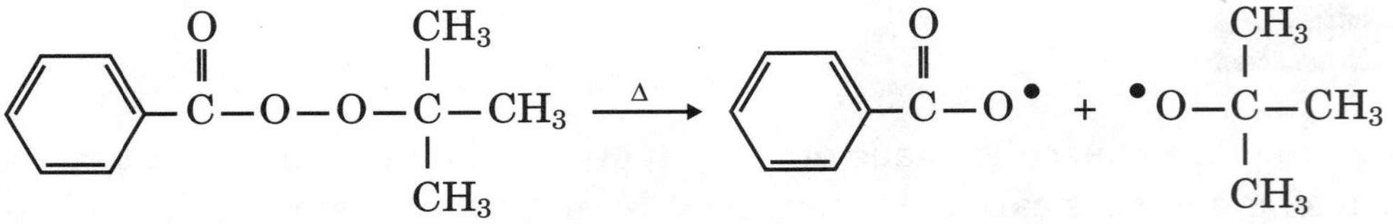

t-butyl benzoyl peroxide

The dot indicates the unpaired electron. The free radical reacts with styrene and initiates the reaction:

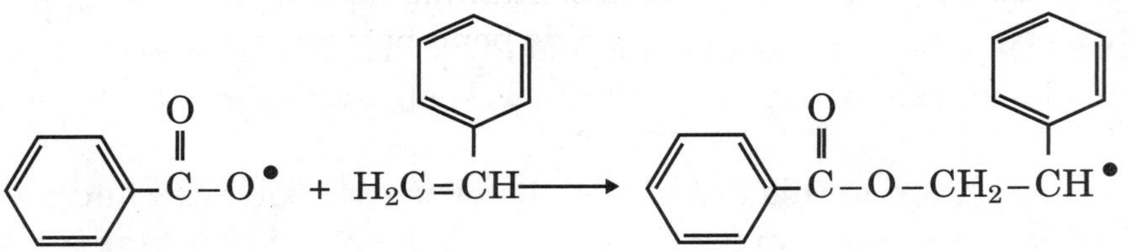

styrene

After this, the styrene monomers are added to the growing chain one by one until giant molecules containing hundreds and thousands of styrene-repeating units are formed. Please note the distinction between the monomer and the repeating unit. The monomer is the starting material, and the repeating unit is part of the polymer chain. Chemically they are not identical. In the case of styrene, the monomer contains a double bond, while the repeating unit (in the brackets in the following structure) does not.

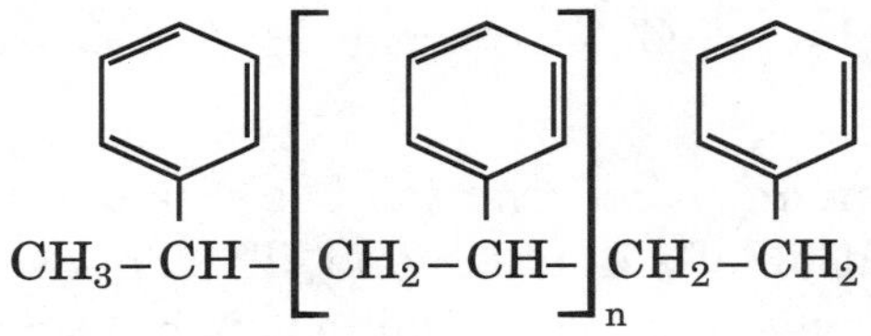

polystyrene

Since the initiators are unstable compounds, care should be taken not to keep them near flames or heat them directly. If a bottle containing a peroxide initiator is dropped, a minor explosion can even occur.

The second type of reaction is called a *condensation reaction* because we condense two monomers into a longer unit, and at the same time we eliminate—expel—a small molecule. This is also called a step growth polymerization reaction. Nylon 6-6 is made of adipoyl chloride and hexamethylene diamine:

$$n \; Cl-\overset{\overset{\displaystyle O}{\|}}{C}-(CH_2)_4-\overset{\overset{\displaystyle O}{\|}}{C}-Cl + n \; H_2N-(CH_2)_6-NH_2 \longrightarrow$$

adipoyl chloride   hexamethylene diamine

$$Cl-\overset{\overset{\displaystyle O}{\|}}{C}-(CH_2)_4-\overset{\overset{\displaystyle O}{\|}}{C}-\left[NH-(CH_2)_6-NH-\overset{\overset{\displaystyle O}{\|}}{C}-(CH_2)_4-\overset{\overset{\displaystyle O}{\|}}{C}-\right]_n NH-(CH_2)_6-NH_2 \; + \; n \; HCl$$

repeating unit

We form an amide linkage between the adipoyl chloride and the amine with the elimination of HCl. The polymer is called nylon 6–6 because there are six carbon atoms in the acyl chloride and six carbon atoms in the diamine. Other nylons, such as nylon 10–6, are made of sebacoyl chloride (a 10-carbon atom containing acyl chloride) and hexamethylene diamine (a six carbon atom containing diamine). We use an acyl chloride rather than a carboxylic acid to form the amide bond because the former is more reactive. NaOH is added to the polymerization reaction in order to neutralize the HCl that is released every time an amide bond is formed.

The length of the polymer chain formed in both reactions depends on environmental conditions. Usually the chains formed can be made longer by heating the products longer. This process is called *curing*.

Objectives

1. To acquaint students with the conceptual and physical distinction between monomer and polymer.
2. To perform addition and condensation polymerization and solvent casting of films.

Procedure

Preparation of Polystyrene

1. Set up your hot plate in the hood. Place 50 g of sea sand in a 150-mL beaker. Position a thermometer (0–200°C) in the sand bath so that it does not touch the bottom of the beaker. Heat the sand bath to 140°C.

2. Place approximately 2.5 mL styrene in a 16/18 × 150-mm Pyrex test tube. Add 3 drops of *t*-butyl benzoyl peroxide (*t*-butyl peroxide benzoate) initiator. Mix the solution.

3. Place the test tube in a test tube holder. Immerse the test tube in the sand bath. **Make sure the test tube points away from your face. (Caution: Do not touch either the test tube or the beaker with your hand.)** Heat the mixture in the test tube to about 140°C. The mixture will turn yellow.

4. When bubbles appear, remove the test tube from the sand bath. The polymerization reaction is exothermic and thus it generates its own heat. Overheating would create sudden boiling. When the bubbles disappear, put the test tube back in the sand bath. But every time the mixture starts to boil you must remove the test tube.

5. Continue the heating until the mixture in the test tube has a syrupy consistency.

6. Immerse a glass rod in the hot mixture. Swirl it around a few times. Remove the glass rod immediately. A chunk of polystyrene will be attached to the glass rod which will solidify upon cooling. The remaining polystyrene will solidify on the walls of the test tube when you remove it from the sand bath. Turn off the hot plate and let the sand bath cool to room temperature. While it is cooling, add a few drops of xylene to the test tube and dissolve some of the polystyrene by warming it in the sand bath.

7. Pour a few drops of the warm xylene solution on a microscopic slide and let the solvent evaporate. A thin film of polystyrene will be obtained. This is one of the techniques—the so-called solvent-casting technique—used to make films from bulk polymers.

8. Discard the remaining xylene solution into a special jar labeled "Waste." Discard the test tube with the polystyrene in a special box labeled "Glass."

9. Investigate the consistency of the solidified polystyrene on your glass rod, removing the solid mass by prying it off with a spatula.

Preparation of Nylon

1. Set up a 50-mL reaction beaker and clamp above it a cylindrical paper roll (from toilet paper) or a stick.

2. Add 2.0 mL of 20% NaOH solution and 10 mL of a 5% aqueous solution of hexamethylene diamine.

3. Take 10 mL of 5% adipoyl chloride solution in cyclohexane with a pipet or syringe. Layer the cyclohexane solution slowly on top of the aqueous solution in the beaker. Two layers will form and nylon will be produced at the interface (Fig. 9.1).

4. With a bent wire first scrape off the nylon formed on the walls of the beaker.

Figure 9.1
Preparation of nylon.

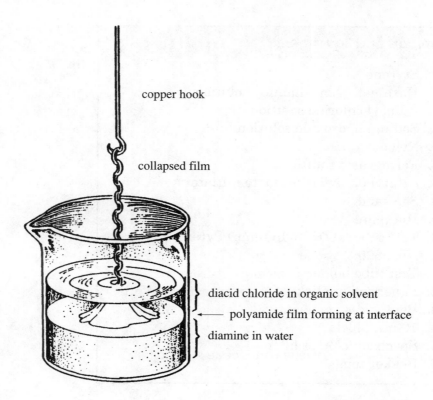

copper hook

collapsed film

diacid chloride in organic solvent
polyamide film forming at interface
diamine in water

5. Slowly lift and pull the film from the center. If you pull it too fast, the nylon rope will break.

6. Wind it around the paper roll or stick two to three times. Do not touch it with your hands.

7. Slowly rotate the roll or the stick and wind at least a 1-m nylon rope.

8. Cut the rope and transfer the wound rope into a beaker filled with water (or 50% ethanol). Watch as the thickness of the rope collapses. Dry the rope between two filter papers.

9. There are still monomers left in the beaker. Mix the contents vigorously with a glass rod. Observe the beads of nylon that have formed.

10. Pour the mixture into a cold water bath and wash it. Dry the nylon between two filter papers. Note the consistency of your products.

11. Dissolve a small amount of nylon in 80% formic acid. Place a few drops of the solution onto a microscope slide and evaporate the solvent under the hood.

12. Compare the appearance of the solvent-cast nylon film with that of the polystyrene.

Chemicals and Equipment

1. Styrene
2. Hexamethylene diamine solution
3. Adipoyl chloride solution
4. Sodium hydroxide solution
5. Xylene
6. Formic acid solution
7. *t*-Butyl peroxide benzoate initiator
8. Sea sand
9. Hot plate
10. Test tube (16/18 × 150-mm) Pyrex No. 9820
11. Test tube holder
12. Paper roll or stick
13. Bent wires
14. 10-mL pipets
15. Spectroline pipet filler
16. Beaker tongs

Experiment 9

PRE-LAB QUESTIONS

1. Why should you *not* expose *t*-butyl peroxide to direct heat?

2. Write the structure of the reaction: *t*-butyl free radical plus styrene yields a *t*-butyl-styrene free radical.

3. Write the reaction for the polymerization of vinyl chloride (chloroethene). Show the repeating unit of the resulting polymer (Polyvinyl chloride, PVC).

4. Write the structure of the monomers and that of the repeating unit in nylon 6-10. (In numbering nylons the first number indicates the number of carbon atoms in the acyl chloride and the second number refers to the number of carbon atoms in the diamine.)

5. Why do we call nylon a condensation polymer?

Experiment 9

REPORT SHEET

1. Describe the appearance of polystyrene and nylon.

2. Describe the difference in physical characteristics between polystyrene and nylon.

3. Is there any difference in the appearance of the solvent cast films of nylon and polystyrene?

POST-LAB QUESTIONS

1. A polyester is made of sebacoyl chloride and ethylene glycol,

$$Cl-\underset{\underset{\displaystyle O}{\parallel}}{C}-(CH_2)_8-\underset{\underset{\displaystyle O}{\parallel}}{C}-Cl \quad \text{and} \quad \underset{\displaystyle CH_2OH}{\overset{\displaystyle CH_2OH}{|}}$$

a. Draw the structure of the polyester formed.

b. What molecules have been eliminated in this condensation reaction?

2. Two compounds, $Cl-\underset{\underset{\displaystyle O}{\parallel}}{C}-(CH_2)_3-\underset{\underset{\displaystyle O}{\parallel}}{C}-Cl$ in cyclohexane and $H_2N-(CH_2)_4-NH_2$ in water, are reacted. Write the structure of the polyamide rope formed.

3. What compound did neutralize the evolving HCl in the preparation of nylon? In what part of the reaction was this supplied?

4. Distinguish between the polarities of the solvents which solubilize polystyrene and nylon, respectively.

Experiment 10

Preparation of acetylsalicylic acid (aspirin)

Background

One of the most widely used nonprescription drugs is aspirin. In the United States, more than 15,000 pounds are sold each year. It is no wonder there is such wide use when one considers the medicinal applications for aspirin. It is an effective analgesic (pain killer) that can reduce the mild pain of headaches, toothache, neuralgia (nerve pain), muscle pain, and joint pain (from arthritis and rheumatism). Aspirin behaves as an antipyretic drug (it reduces fever) and an antiinflammatory agent capable of reducing the swelling and redness associated with inflammation. It is an effective agent in preventing strokes and heart attacks due to its ability to act as an anticoagulant by preventing platelet aggregation.

Early studies showed the active agent that gave these properties to be salicylic acid. However, salicylic acid contains the phenolic and the carboxylic acid groups. As a result, the compound was too harsh to the linings of the mouth, esophagus, and stomach. Contact with the stomach lining caused some hemorrhaging. The Bayer Company in Germany patented the ester acetylsalicylic acid and marketed the product as "aspirin" in 1899. Their studies showed that this material was less of an irritant; the acetylsalicylic acid was hydrolyzed in the small intestine to salicylic acid, which then was absorbed into the bloodstream. The relationship between salicylic acid and aspirin is shown in the following formulas:

Salicylic acid

Acetylsalicylic acid (Aspirin)

Aspirin still has side effects. Hemorrhaging of the stomach walls can occur even with normal dosages. These side effects can be reduced through the addition of coatings or through the use of buffering agents. Magnesium hydroxide, magnesium carbonate, and aluminum glycinate, when mixed into the formulation of the aspirin (e.g., Bufferin), reduce the irritation.

This experiment will acquaint you with a simple synthetic problem in the preparation of aspirin. The preparative method uses acetic anhydride and an acid catalyst, like sulfuric or phosphoric acid, to speed up the reaction with salicylic acid.

| Salicylic acid | Acetic anhydride | | Aspirin | Acetic acid |

If any salicylic acid remains unreacted, its presence can be detected with a 1% iron(III) chloride solution. Salicylic acid has a phenol group in the molecule. The iron(III) chloride gives a violet color with any molecule possessing a phenol group (see Experiment 30). Notice the aspirin no longer has the phenol group. Thus a pure sample of aspirin will not give a purple color with 1% iron(III) chloride solution.

Objectives

1. To illustrate the synthesis of the drug aspirin.
2. To use a chemical test to determine the purity of the preparation.

Procedure

Preparation of Aspirin

1. Prepare a bath using a 400-mL beaker filled about half way with water. Heat to boiling.

2. Take 2.0 g of salicylic acid and place it in a 125-mL Erlenmeyer flask. Use this quantity of salicylic acid to calculate the theoretical or expected yield of aspirin (1). Carefully add 3 mL of acetic anhydride to the flask and, while swirling, add 3 drops of concentrated phosphoric acid.

CAUTION!

Acetic anhydride will irritate your eyes. Phosphoric acid will cause burns to the skin. Use gloves with these reagents. Handle both chemicals with care. Dispense in the hood.

3. Mix the reagents and then place the flask in the boiling water bath; heat for 15 min. (Fig. 10.1). The solid will completely dissolve. Swirl the solution occasionally.

Figure 10.1
Assembly for the
synthesis of aspirin.

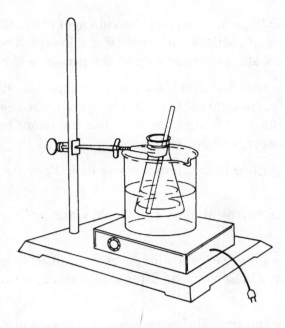

4. Remove the Erlenmeyer flask from the bath and let it cool to approximately room temperature. Then, slowly pour the solution into a 150-mL beaker containing 20 mL of ice water, mix thoroughly, and place the beaker in an ice bath. The water destroys any unreacted acetic anhydride and will cause the insoluble aspirin to precipitate from solution.

5. Collect the crystals by filtering under suction with a Büchner funnel. The assembly is shown in Fig. 10.2. (Also see Fig. 6.1, p. 75.)

Figure 10.2
Filtering using the
Büchner funnel.

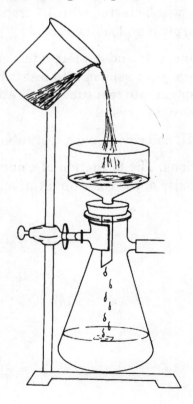

6. Obtain a 250-mL filter flask and connect the side arm of the filter flask to a water aspirator with heavy wall vacuum rubber tubing. (The thick walls of the tubing will not collapse when the water is turned on and the pressure is reduced.)

7. The Büchner funnel is inserted into the filter flask through either a filtervac, a neoprene adapter, or a one-hole rubber stopper, whichever is available. Filter paper is then placed into the Büchner funnel. Be sure that the paper lies flat and covers all the holes. Wet the filter paper with water.

8. Turn on the water aspirator to maximum water flow. Pour the solution into the Büchner funnel.

9. Wash the crystals with two 5-mL portions of cold water, followed by one 10-mL portion of cold ethanol.

10. Continue suction through the crystals for several minutes to help dry them. Disconnect the rubber tubing from the filter flask before turning off the water aspirator.

11. Using a spatula, place the crystals between several sheets of paper toweling or filter paper and press dry the solid.

12. Weigh a 50-mL beaker (2). Add the crystals and reweigh (3). Calculate the weight of crude aspirin (4). Determine the percent yield (5).

Determine the Purity of the Aspirin

1. The aspirin you prepared is not pure enough for use as a drug and is *not* suitable for ingestion. The purity of the sample will be tested with 1% iron(III) chloride solution and compared with a commercial aspirin and salicylic acid.

2. Label three test tubes (100 × 13 mm) 1, 2, and 3; place a few crystals of salicylic acid into test tube no. 1, a small sample of your aspirin into test tube no. 2, and a small sample of a crushed commercial aspirin into test tube no. 3. Add 5 mL of distilled water to each test tube and swirl to dissolve the crystals.

3. Add 10 drops of 1% aqueous iron(III) chloride to each test tube.

4. Compare and record your observations. The formation of a purple color indicates the presence of salicylic acid. The intensity of the color qualitatively tells how much salicylic acid is present.

Chemicals and Equipment

1. Acetic anhydride
2. Concentrated phosphoric acid, H_3PO_4
3. Commercial aspirin tablets
4. 95% Ethanol
5. 1% Iron(III) chloride
6. Salicylic acid
7. Boiling chips
8. Büchner funnel, small
9. 250-mL filter flask
10. Filter paper
11. Filtervac or neoprene adapter
12. Hot plate

Experiment  11

Measurement of the active ingredient in aspirin pills

Background

Medication delivered in the form of a pill contains an active ingredient or ingredients. Beside the drug itself, the pill also contains fillers. The task of the filler is many fold. Sometimes it is there to mask the bitter or otherwise unpleasant taste of the drug. Other times the filler is necessary because the prescribed dose of the drug is so small in mass that it would be difficult to handle. Drugs that have the same generic name contain the same active ingredient. The dosage of the active ingredient must be listed as specified by law. On the other hand, neither the quantity of the filler nor its chemical nature appears on the label. That does not mean that the fillers are completely inactive. They usually affect the rate of drug delivery. In order to deliver the active ingredient, the pill must fall apart in the stomach. For this reason, many fillers are polysaccharides, for example, starch, that either are partially soluble in stomach acid or swell, allowing the drug to be delivered in the stomach or in the intestines.

In the present experiment, we measure the amount of the active ingredient, acetylsalicylic acid (see also Experiment 10), in common aspirin pills. Companies use different fillers and in different amounts, but the active ingredient, acetylsalicylic acid, must be the same in every aspirin tablet. We separate the acetylsalicylic acid from the filler based on their different solubilities. Acetylsalicylic acid is very soluble in ethanol, while neither starch, nor other polysaccharides, or even mono- and disaccharides used as a fillers, are soluble in ethanol. Some companies may use inorganic salts as fillers but these too are not soluble in ethanol. On the other hand, some specially formulated aspirin tablets may contain small amounts of ethanol-soluble substances such as stearic acid or vegetable oil. Thus the ethanol extracts of aspirin tablets may contain small amounts of substances other than acetylsalicylic acid.

Objectives

1. To appreciate the ratio of filler to active ingredients in common aspirin tablets.
2. To learn techniques of quantitative separations.

Procedure

1. Weigh approximately 10 g of aspirin tablets. Record the actual weight on your Report Sheet (1). Count the number of tablets and record it on your Report Sheet (2).

2. Place the weighed aspirin tablets in a mortar of approximately 100 mL capacity. Before starting to grind, place the mortar on a white sheet of paper and loosely cover it with a filter paper. The purpose of this procedure is to catch small fragments of the tablets that may fly out of the mortar during the grinding process. Break up the aspirin tablets by gently hammering them with the pestle. Recover and place back in the mortar any fragments that flew out during the hammering. With a twisting motion of your wrist, grind the aspirin pieces into a fine powder with the aid of the pestle.

3. Add 10 mL of 95% ethanol to the mortar and continue to grind for 2 min. Place a filter paper (Whatmann no. 2, 7 cm) in a funnel and place the funnel in a 250-mL Erlenmeyer flask. With the aid of a glass rod, transfer the supernatant liquid from the mortar to the filter paper. After a few minutes, when about 1 mL of clear filtrate has been collected in the Erlenmeyer flask, lift the funnel and allow a drop of the filtrate to fall on a clean microscope slide. Replace the funnel in the Erlenmeyer flask and allow the filtration to continue. The drop on the microscope slide will rapidly evaporate leaving behind crystals of acetylsalicylic acid. This is a qualitative test showing that the extraction of the active ingredient is successful. Report what you see on the microscope slide on your Report Sheet (3).

4. Add another 10 mL of 95% ethanol and repeat the procedure from no. 3.

5. Repeat procedure no. 4 two more times; you will use a total of 40 mL of ethanol in the four extractions. Report after each extraction if the extract carries acetylsalicylic acid. Enter these observations on your Report Sheet (4, 5, and 6).

6. When the filtration is completed and only the white, moist solid is left in the filter, transfer the filter paper with its contents into a 100-mL beaker and place the beaker into a drying oven set at 110°C. Dry for 10 min.

7. Carefully remove the beaker from the oven. (**CAUTION!** The beaker is hot.) Allow it to come to room temperature. Weigh a clean and dry 25-mL beaker on your balance. Report the weight on your Report Sheet (7). With the aid of a spatula, **carefully** transfer the dried filler from the filter paper into the 25-mL beaker. Make sure that you do not spill any of the powder. Some of the dried filler may stick to the paper a bit, and you may have to scrape the paper with the spatula. Weigh the 25-mL beaker with its contents on your balance. Report the weight on your Report Sheet (8).

8. Test the dried filler with a drop of Hanus iodine solution. A blue coloration will indicate that it contains starch. Report your findings on your Report Sheet (14).

Chemicals and Equipment

1. Aspirin tablets
2. Mortar and pestle (100-mL capacity)
3. 95% ethanol
4. Filter paper (Whatman no. 2, 7 cm)
5. Balance
6. Drying oven at 110°C
7. Hanus iodine solution
8. Microscope slides
9. 100-mL beaker
10. 25-ml beaker

EXPERIMENT 11

PRE-LAB QUESTIONS

1. What method is used to separate acetylsalicylic acid from starch?

2. What is the role of a filler, like starch, in influencing the effect of aspirin?

3. The normal adult aspirin tablet contains 5.4 grains of aspirin. If 1 grain is 64.8 mg, how many milligrams of aspirin are in one tablet?

Experiment 11

REPORT SHEET

1. Weight of aspirin tablets _____ g

2. Number of aspirin tablets in your sample _____

3. Does your first extract contain acetylsalicylic acid? _____

4. Does your second extract contain acetylsalicylic acid? _____

5. Does your third extract contain acetylsalicylic acid? _____

6. Does your fourth extract contain acetylsalicylic acid? _____

7. Weight of the empty 25-mL beaker _____ g

8. Weight of the 25-mL beaker and filler _____ g

9. Weight of the filler: (8) − (7) _____ g

10. Percent of filler in tablets: [(9)/(1)] × 100 = % _____ %

11. Weight of one tablet: (1)/(2) _____ g

12. Weight of filler per tablet: (11) × [(10)/100] _____ g

13. Weight of acetylsalicylic acid per tablet: (11) − (12) _____ g

14. Does your filler contain starch? _____

POST-LAB QUESTIONS

1. According to your calculations, does your aspirin tablet contain more, the same, or less active ingredients than the average adult dosage (5.4 grains)?

2. If your ethanol extract contained a filler in addition to the active ingredient, acetylsalicylic acid, how would that affect your calculations of the dosage of the aspirin tablet?

3. If instead of starch the filler would be inorganic salt, would your procedure yield the same, correct aspirin content?

4. On the basis of the Hanus iodine test performed, what can you say about the nature of the filler in your aspirin tablet?

5. If you did not properly grind your aspirin tablet to a fine powder, would you need more, less, or an equal amount of ethanol extraction to remove most of the aspirin?

6. You obtained a painkiller pill containing acetaminophen. After extracting the active ingredient from a 400-mg pill, you ended up with the following data: beaker: 5.38 g; beaker plus filler: 5.66 g. What was the percent of filler in the painkiller tablet?

Harcourt, Inc.

Isolation of caffeine from tea leaves

Background

Many organic compounds are obtained from natural sources through extraction. This method takes advantage of the solubility characteristics of a particular organic substance with a given solvent. In the experiment here, caffeine is readily soluble in hot water and is thus separated from the tea leaves. Caffeine is one of the main substances that make up the water solution called tea. Besides being found in tea leaves, caffeine is present in coffee, kola nuts, and cocoa beans. As much as 5% by weight of the leaf material in tea plants consists of caffeine.

The caffeine structure is shown below. It is classed as an alkaloid, meaning that with the nitrogen present, the molecule has base characteristics (alkali-like). In addition, the molecule has the purine ring system, a framework which plays an important role in living systems.

Caffeine is the most widely used of all the stimulants. Small doses of this chemical (50 to 200 mg) can increase alertness and reduce drowsiness and fatigue. The "No-Doz" tablet lists caffeine as the main ingredient. In addition, it affects blood circulation since the heart is stimulated and blood vessels are relaxed (vasodilation). It also acts as a diuretic. There are side effects. Large doses of over 200 mg can result in insomnia, restlessness, headaches, and muscle tremors ("coffee nerves"). Continued, heavy use may bring on physical dependence. (How many of you know somebody who cannot function in the morning until they have that first cup of coffee?)

Tea leaves consist primarily of cellulose; this is the principle structural material of all plant cells. Fortunately, the cellulose is insoluble in water, so that by using a hot water extraction, more soluble caffeine can be separated. Also dissolved in water are complex substances called tannins. These are colored phenolic compounds of high molecular weight (500 to 3000) that have acidic behavior. If a basic salt such as Na_2CO_3 is added to the water solution, the tannins can react to form a salt. These salts are insoluble in organic solvents, such as chloroform or dichloromethane, but are soluble in water.

Although caffeine is soluble in water (2 g/100 g of cold water), it is more soluble in the organic solvent dichloromethane (14 g/100 g). Thus caffeine can be extracted from the basic tea solution with dichloromethane, but the sodium salts of the tannins remain behind in the aqueous solution. Evaporation of the dichloromethane yields crude caffeine; the crude material can be purified by sublimation.

Objectives

1. To demonstrate the isolation of a natural product.
2. To learn the techniques of extraction.
3. To use sublimation as a purification technique.

Procedure

The isolation of caffeine from tea leaves follows the scheme below:

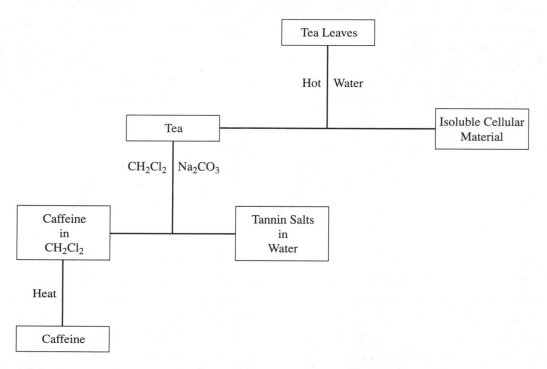

1. Carefully open two commercial tea bags (try not to tear the paper) and weigh the contents to the nearest 0.001 g. Record this weight (1). Place the tea leaves back into the bags, close, and secure the bags with staples.

2. Into a 150-mL beaker, place the tea bags so that they lie flat on the bottom. Add 30 mL of distilled water and 2.0 g of anhydrous Na_2CO_3; heat the contents with a hot plate, keeping a *gentle* boil, for 20 min. While the mixture is boiling, keep a watch glass on the beaker. Hold the tea bags under water by occasionally pushing them down with a glass rod.

3. Decant the hot liquid into a 50-mL Erlenmeyer flask. Wash the tea bags with 10 mL of hot water, carefully pressing the tea bag with a glass rod; add this wash water to the tea

extract. (If any solids are present in the tea extract, filter them by gravity to remove.) Cool the combined tea extract to room temperature. The tea bags may be discarded.

4. Transfer the cool tea extract to a 125-mL separatory funnel that is supported on a ring stand with a ring clamp.

5. Carefully add 5.0 mL of dichloromethane to the separatory funnel. Stopper the funnel and lift it from the ring clamp; hold the funnel with two hands as shown in Fig. 12.1. By holding the stopper in place with one hand, invert the funnel. *Make certain the stopper is held tightly and no liquid is spilled; make sure the liquid is not in contact with the stop-cock; open the stop-cock, being sure to point the opening away from you and your neighbors.* Built-up pressure caused by gases accumulating inside will be released. Now, close the stop-cock and gently mix the contents by inverting the funnel two or three times. Again, release any pressure by opening the stop-cock as before.

Figure 12.1
Using the separatory funnel.

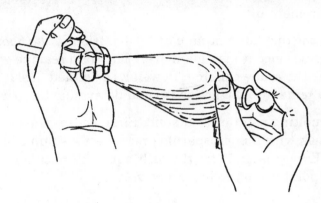

6. Return the separatory funnel to the ring clamp, remove the stopper, and allow the aqueous layer to separate from the dichloromethane layer (Fig. 12.2). You should see two distinct layers form after a few minutes, with the dichloromethane layer at the bottom. Sometimes an emulsion may form at the juncture of the two layers. The emulsion often can be broken by gently swirling the contents or by gently stirring the emulsion with a glass rod.

Figure 12.2
Separation of the aqueous layer and the dichloromethane layer in the separatory funnel.

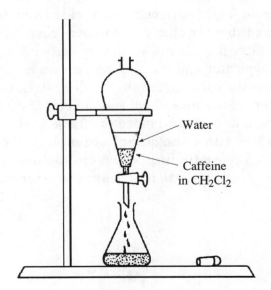

Water

Caffeine in CH_2Cl_2

7. Carefully drain the lower layer into a 25-mL Erlenmeyer flask. Try not to include any water with the dichloromethane layer; careful manipulation of the stop-cock will prevent this.

8. Repeat the extraction with an additional 5.0 mL of dichloromethane. Combine the separated bottom layer with the dichloromethane layer obtained from step no. 7.

9. Add 0.5 g of anhydrous Na_2SO_4 to the combined dichloromethane extracts. Swirl the flask. The anhydrous salt is a drying agent and will remove any water that may still be present.

10. Weigh a 25-mL side-arm filter flask containing one or two boiling stones. Record this weight (2). By means of a gravity filtration, filter the dichloromethane–salt mixture into the pre-weighed flask. Rinse the salt on the filter paper with an additional 2.0 mL of dichloromethane.

11. Remove the dichloromethane by evaporation *in the hood*. Be careful not to overheat the solvent, since it may foam over. The solid residue which remains after the solvent is gone is the crude caffeine. Reweigh the cooled flask (3). Calculate the weight of the crude caffeine by subtraction (4) and determine the percent yield (5).

12. Take a melting point of your solid. First, scrape the caffeine from the bottom and sides of the flask with a microspatula and collect a sample of the solid in a capillary tube (review Experiment 15 for the technique). Pure caffeine melts at 238°C. Compare your melting point (6) to the literature value.

Optional

13. At the option of your instructor, the caffeine may be purified further. The caffeine may be sublimed directly from the flask with a cold finger condenser (Fig. 12.3). Carefully insert the cold finger condenser into a no. 2 neoprene adapter (use a drop of glycerine as a lubricant). Adjust the tip of the cold finger to 1 cm from the bottom of the flask. Clean any glycerine remaining on the cold finger with a Kimwipe and acetone; the cold finger surface must be clean and dry. Connect the cold finger to a faucet by latex tubing (water *in* the upper tube; water *out* the lower tube). Connect the side-arm filter flask to a water aspirator with vacuum tubing, installing a trap between the aspirator and the sublimation set-up (Fig. 12.3). When you turn the water on, press the cold finger into the filter flask until a good seal is made. Gently heat the bottom of the filter flask which holds the caffeine with a microburner (hold the *base* of the microburner); move the flame back and forth and along the sides of the flask. **Do not allow the sample to melt.** If the sample melts, *stop* heating and allow to cool before continuing. When the sublimation is complete, disconnect the heat and allow the system to cool; leave the aspirator connected and the water running.

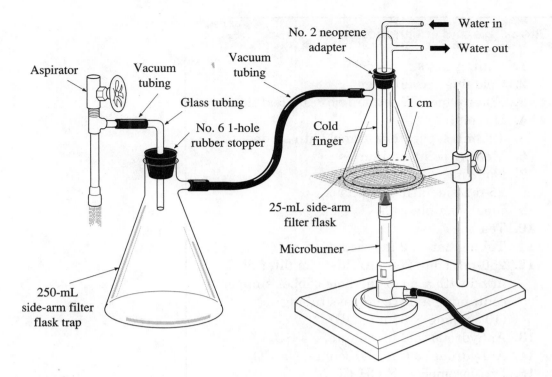

Figure 12.3 • Sublimation apparatus connected to an aspirator.

14. When the system has reached room temperature, carefully disconnect the aspirator from the side-arm filter flask by removing the vacuum tubing from the side-arm. Turn off the water to the cold finger. *Carefully* remove the cold finger from the flask along with the neoprene adapter without dislodging any crystals. Scrape the sublimed caffeine onto a pre-weighed piece of weighing paper (7). Reweigh (8); determine the weight of caffeine (9). Calculate the percent recovery (10). Determine the melting point (11).

15. Collect the caffeine in a sample vial, and submit it to your instructor.

Chemicals and Equipment

1. Boiling chips
2. Cold finger condenser
3. Filter paper (Whatman no. 7.0), fast flow
4. Hot plate
5. 125-mL separatory funnel with stopper
6. Melting point capillaries
7. No. 2 neoprene adapter
8. 25-mL side-arm filter flask
9. Small sample vials
10. Tea bags
11. Tubing: latex, 2 ft.; vacuum, 2 ft.
12. 250-mL trap: 250-mL side-arm filter flask fitted with a no. 6 one-hole rubber stopper containing a piece of glass tubing (10 cm long $\times$ 37 mm OD)
13. Anhydrous sodium sulfate, Na_2SO_4
14. Anhydrous sodium carbonate, Na_2CO_3
15. Dichloromethane, CH_2Cl_2
16. Stapler

Harcourt, Inc.

Experiment 12

PRE-LAB QUESTIONS

1. What method is used to obtain caffeine for tea leaves?

2. Why is caffeine classed as an alkaloid?

3. Why might an individual use a product containing caffeine?

4. Besides caffeine, what other compounds are found in tea leaves?

Experiment 12

REPORT SHEET

1. Weight of tea in 2 tea bags _____ g

2. Weight of 25-mL side-arm filter flask and boiling stones _____ g

3. Weight of flask, boiling stones, and crude caffeine _____ g

4. Weight of caffeine: (3) − (2) _____ g

5. Percent yield: [(4)/(1)] × 100 = % _____ %

6. Melting point of your crude caffeine _____ °C

7. Weight of weighing paper _____ g

8. Weight of sublimed caffeine and paper _____ g

9. Weight of caffeine: (8) − (7) _____ g

10. Percent recovery: [(9)/(4)] × 100 = % _____ %

11. Melting point of sublimed caffeine _____ °C

POST-LAB QUESTIONS

1. A student used 5.326 g of tea leaves in the experiment. How much caffeine is expected, assuming all the caffeine is extracted? (*Hint:* see **Background.**)

2. How is the purity of the recovered caffeine determined?

3. In the isolation procedure (step no. 2), sodium carbonate, Na_2CO_3, is added to the water solution. Explain why this is done.

4. What did the use of anhydrous sodium sulfate accomplish?

Experiment 13

Carbohydrates

Background

Carbohydrates are polyhydroxy aldehydes, ketones, or compounds that yield polyhydroxy aldehydes or ketones upon hydrolysis. Rice, potatoes, bread, corn, candy, and fruits are rich in carbohydrates. A carbohydrate can be classified as a monosaccharide (glucose or fructose); a disaccharide (sucrose or lactose), which consists of two joined monosaccharides; or a polysaccharide (starch or cellulose), which consists of thousands of monosaccharide units linked together. Monosaccharides exist mostly as cyclic structures containing hemiacetal (or hemiketal) groups. These structures in solutions are in equilibrium with the corresponding open chain structures bearing aldehyde or ketone groups. Glucose, blood sugar, is an example of a polyhydroxy aldehyde (Fig. 13.1).

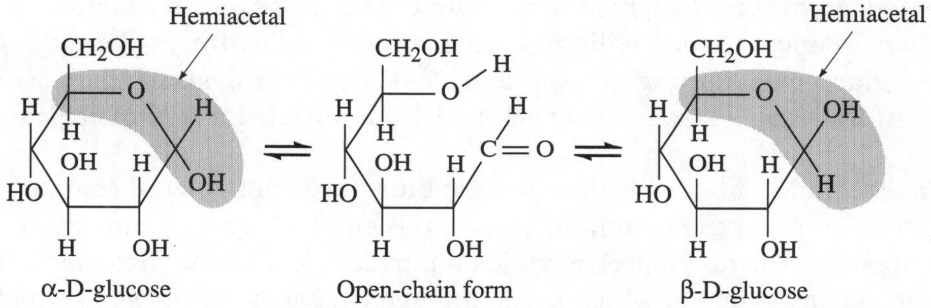

Figure 13.1 • The structures of D-glucose.

Disaccharides and polysaccharides exist as cyclic structures containing functional groups such as hydroxyl groups, acetal (or ketal), and hemiacetal (or hemiketal). Most of the di-, oligo-, or polysaccharides have two distinct ends. The one end which has a hemiacetal (or hemiketal) on its terminal is called the reducing end, and the one which does not contain a hemiacetal (or hemiketal) terminal is the nonreducing end. The name "reducing" is given because hemiacetals (and to a lesser extent hemiketals) can reduce an oxidizing agent such as Benedict's reagent.

Fig. 13.2 is an example:

Figure 13.2
The structure of maltose, a disaccharide.

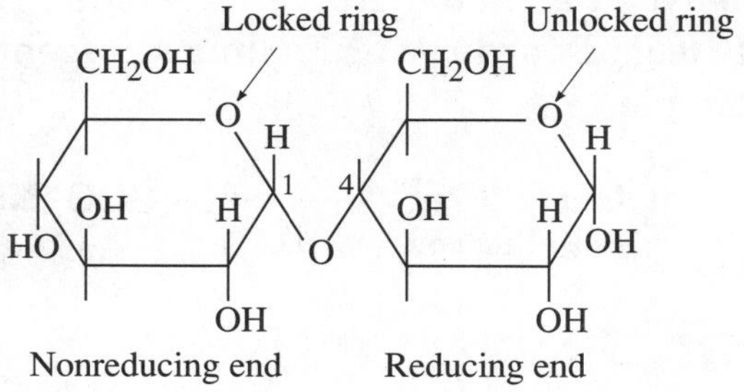

Not all disaccharides or polysaccharides contain a reducing end. An example is sucrose, which does not have a hemiacetal (or hemiketal) group on either of its ends (Fig. 13.3).

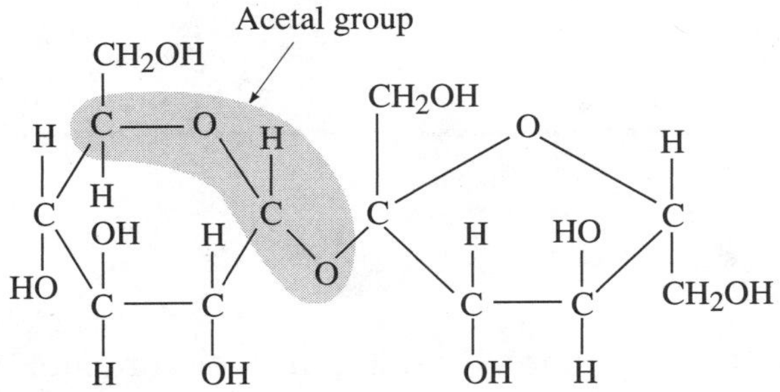

Figure 13.3
The structure of sucrose.

Polysaccharides, such as amylose or amylopectin, do have a hemiacetal group on one of their terminal ends, but practically they are nonreducing substances because there is only one reducing group for every 2,000–10,000 monosaccharidic units. In such a low concentration, the reducing group does not give a positive test with Benedict's or Fehling's reagent.

On the other hand, when a nonreducing disaccharide (sucrose) or a polysaccharide such as amylose is hydrolyzed the glycosidic linkages (acetal) are broken and reducing ends are created. Hydrolyzed sucrose (a mixture of D-glucose and D-fructose) will give a positive test with Benedict's or Fehling's reagent as well as hydrolyzed amylose (a mixture of glucose and glucose containing oligosaccharides). The hydrolysis of sucrose or amylose can be achieved by using a strong acid such as HCl or with the aid of biological catalysts (enzymes).

Starch can form an intense, brilliant, dark blue-, or violet-colored complex with iodine. The straight chain component of starch, the amylose, gives a blue color while the branched component, the amylopectin, yields a purple color. In the presence of iodine, the amylose forms helixes inside of which the iodine molecules assemble as long polyiodide chains. The helix-forming branches of amylopectin are much shorter than those of amylose. Therefore, the polyiodide chains are also much shorter in the amylopectin-iodine complex than in the amylose-iodine complex. The result is a different color (purple). When starch is hydrolyzed and broken down to small carbohydrate units, the iodine will not give a dark blue (or purple) color. The iodine test is used in this experiment to indicate the completion of the hydrolysis.

In this experiment, you will investigate some chemical properties of carbohydrates in terms of their functional groups.

1. *Reducing and nonreducing properties of carbohydrates*

   a. **Aldoses (polyhydroxy aldehydes).** All aldoses are reducing sugars because they contain free aldehyde functional groups. The aldehydes are oxidized by mild oxidizing agents (e.g., Benedict's or Fehling's reagent) to the corresponding carboxylates. For example,

$$R—CHO + 2Cu^{2+} \xrightarrow{\text{NaOH}} R—COO^-Na^+ + Cu_2O \downarrow$$

(from Fehling's reagent)　　　　　　Red precipitate

b. Ketoses (polyhydroxy ketones). All ketoses are reducing sugars because they have a ketone functional group next to an alcohol functional group. The reactivity of this specific ketone (also called α-hydroxyketone) is attributed to its ability to form an α-hydroxyaldehyde in basic media according to the following equilibrium equations:

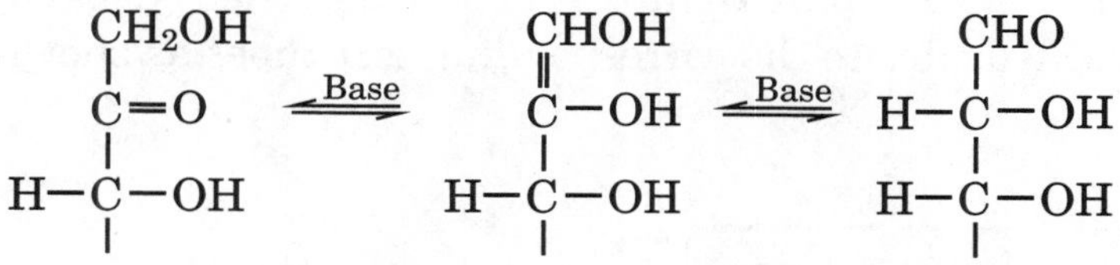

| Ketose | Enediol | Aldose |

c. Hemiacetal functional group (potential aldehydes). Carbohydrates with hemiacetal functional groups can reduce mild oxidizing agents such as Benedict's reagent because hemiacetals can easily form aldehydes through the following equilibrium equation:

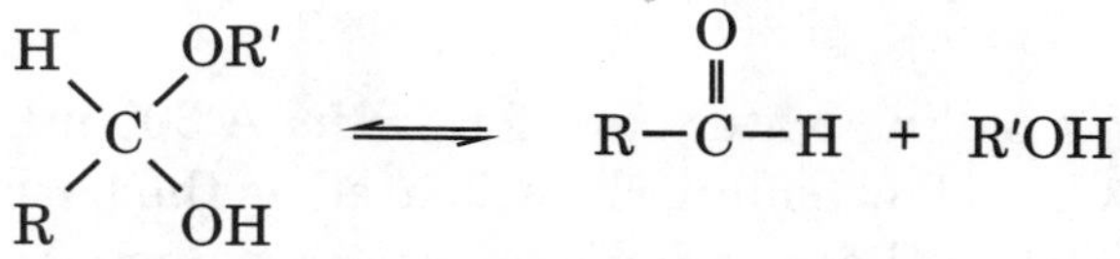

Sucrose is, on the other hand, a nonreducing sugar because it does not contain a hemiacetal functional group. Although starch has a hemiacetal functional group at one end of its molecule, it is, however, considered as a nonreducing sugar because the effect of the hemiacetal group in a very large starch molecule becomes insignificant to give a positive Benedict's test.

2. *Hydrolysis of acetal groups.* Disaccharides and polysaccharides can be converted into monosaccharides by hydrolysis. The following is an example:

$$C_{12}H_{22}O_{11} + H_2O \xrightarrow{\text{catalyst}} C_6H_{12}O_6 + C_6H_{12}O_6$$

Lactose   Glucose   Galactose
(milk sugar)

Objectives

1. To become familiar with the reducing or nonreducing nature of carbohydrates.
2. To experience the enzyme-catalyzed and acid-catalyzed hydrolysis of acetal groups.

Reducing or Nonreducing Carbohydrates

Place approximately 2 mL (approximately 40 drops) of Fehling's solution (20 drops each of solution part A and solution part B) into each of five labeled tubes. Add 10 drops of each of the following carbohydrates to the corresponding test tubes as shown in the following table.

Test tube no.	Name of carbohydrate
1	Glucose
2	Fructose
3	Sucrose
4	Lactose
5	Starch

Place the test tubes in a boiling water bath for 5 min. A 600-mL beaker containing about 200 mL of tap water with a few boiling chips is used as the bath. Record your results on your Report Sheet. Which of those carbohydrates are reducing carbohydrates?

Hydrolysis of Carbohydrates

Hydrolysis of sucrose (acid versus base catalysis)

Place 3 mL of 2% sucrose solution in each of two labeled test tubes. To the first test tube (no. 1), add 3 mL of water and 3 drops of dilute sulfuric acid solution (3 M H_2SO_4). To the second test tube (no. 2), add 3 mL of water and 3 drops of dilute sodium hydroxide solution (3 M NaOH). Heat the test tubes in a boiling water bath for about 5 min. Cool both solutions to room temperature. To the contents of test tube no. 1, add dilute sodium hydroxide solution (3 M NaOH) (about 10 drops) until red litmus paper turns blue. Test a few drops of each of the two solutions (test tube nos. 1 and 2) with Fehling's reagent as described before. Record your results on your Report Sheet.

Hydrolysis of starch (enzyme versus acid catalysis)

Place 2 mL of 2% starch solution in each of two labeled test tubes. To the first test tube (no. 1), add 2 mL of your own saliva. (Use a 10-mL graduated cylinder to collect your saliva.) To the second test tube (no. 2), add 2 mL of dilute sulfuric acid (3 M H_2SO_4). Place both test tubes in a water bath that has been previously heated to 45°C. Allow the test tubes with their contents to stand in the warm water bath for 30 min. Transfer a few drops of each solution into separate depressions of a spot plate or two separately labeled microtest tubes. (Use two clean, separate medicine droppers for transferring.) To each sample (in microtest tubes or on a spot plate), add 2 drops of iodine solution. Record the color of the solutions on your Report Sheet.

Acid catalyzed hydrolysis of starch

Place 5.0 mL of starch solution in a 15 × 150 mm test tube and add 1.0 mL of dilute sulfuric acid (3 M H_2SO_4). Mix it by gently shaking the test tube. Heat the solution in a boiling water bath for about 5 min. Using a clean medicine dropper, transfer about 3 drops of the starch solution into a spot plate or a microtest tube and then add 2 drops of iodine solution. Observe the color of the solution. If the solution gives a positive test with iodine solution (the solution should turn blue), continue heating. Transfer about 3 drops of the boiling solution at 5-min. intervals for an iodine test. (*Note: Rinse the medicine dropper very thoroughly before each test.*) When the solution no longer gives a blue color with iodine solution, stop heating and record the time needed for the completion of hydrolysis.

> ### Chemicals and Equipment
>
> 1. Bunsen burner
> 2. Medicine droppers
> 3. Microtest tubes or a white spot plate
> 4. Boiling chips
> 5. Fehling's reagent
> 6. 3 M NaOH
> 7. 2% starch solution
> 8. 2% sucrose
> 9. 2% fructose
> 10. 2% glucose
> 11. 2% lactose
> 12. 3 M H_2SO_4
> 13. 0.01 M iodine in KI

Experiment 13

PRE-LAB QUESTIONS

1. Circle and label the hemiacetal functional group and the acetal functional group in the following carbohydrates:

a. sucrose

b. lactose

2. Sucrose is a nonreducing sugar. After complete acid hydrolysis, will there be reducing groups? How many per sucrose molecule?

3. When a reducing sugar reacts with Fehling's reagent, what will be the product besides Cu_2O?

Experiment 13

REPORT SHEET

Reducing or nonreducing carbohydrates

Test tube no.	Substance	Reducing or nonreducing carbohydrates
1	Glucose	
2	Fructose	
3	Sucrose	
4	Lactose	
5	Starch	

Hydrolysis of carbohydrates

Hydrolysis of sucrose (acid versus base catalysis)		
Sample	Condition of hydrolysis	Fehling's reagent (positive or negative)
1	Acidic (H_2SO_4)	
2	Basic (NaOH)	

Hydrolysis of starch (enzyme versus acid catalysis)		
Sample	Condition of hydrolysis	Iodine test (positive or negative)
1	Enzymatic (saliva)	
2	Acidic (H_2SO_4)	

Acid catalyzed hydrolysis of starch		
Test tube no.	Heating time (min.)	Iodine test (positive or negative)
1	5	
2	10	
3	15	
4	20	

POST-LAB QUESTIONS

1. An amylose solution is colorless. The iodine solution is reddish-brown. Yet when you combine these two solutions, you observe an intense blue color. What changes in molecular structures give this coloration?

2. The hydrolysis of starch was stopped when the iodine test no longer gave a blue color. Does this mean that the starch solution was completely hydrolyzed to glucose? Explain.

3. Which hydrolysis of the starch is faster? On the basis of this experiment estimate what will happen to the digestion of a piece of bread (containing starch) when you chew it thoroughly?

4. In an unusual disaccharide, two α-D-glucose units are linked together in an $\alpha(1 \rightarrow 1)$ glycosidic linkage. Is this a reducing or nonreducing disaccharide? Explain.

Experiment 14

Preparation and properties of a soap

Background

A soap is the sodium or potassium salt of a long-chain fatty acid. The fatty acid usually contains 12 to 18 carbon atoms. Solid soaps usually consist of sodium salts of fatty acids, whereas liquid soaps consist of the potassium salts of fatty acids.

A soap such as sodium stearate consists of a nonpolar end (the hydrocarbon chain of the fatty acid) and a polar end (the ionic carboxylate).

$$CH_3CH_2CH_2CH_2CH_2CH_2CH_2CH_2CH_2CH_2CH_2CH_2CH_2CH_2CH_2CH_2CH_2 - \overset{\overset{\textstyle O}{\|}}{C} - O^-Na^+$$

<table>
<tr><td>Nonpolar
(Dissolves in oils)</td><td>Polar
(Dissolves in water)</td></tr>
</table>

Because "like dissolves like," the nonpolar end (hydrophobic or water-hating part) of the soap molecule can dissolve the greasy dirt, and the polar or ionic end (hydrophilic or water-loving part) of the molecule is attracted to water molecules. Therefore, the dirt from the surface being cleaned will be pulled away and suspended in water. Thus soap acts as an *emulsifying agent*, a substance used to disperse one liquid (oil molecules) in the form of finely suspended particles or droplets in another liquid (water molecules).

Treatment of fats or oils with strong bases such as lye (NaOH) or potash (KOH) causes them to undergo hydrolysis (saponification) to form glycerol and the salt of a long-chain fatty acid (soap).

$$
\begin{array}{l}
CH_2 - O - \overset{\overset{\textstyle O}{\|}}{C} - C_{17}H_{35} \\
CH - O - \overset{\overset{\textstyle O}{\|}}{C} - C_{17}H_{35} \quad + \quad 3NaOH \quad \xrightarrow{\Delta} \quad CHOH \quad + \quad 3C_{17}H_{35}\overset{\overset{\textstyle O}{\|}}{C} - O^-Na^+ \\
CH_2 - O - \overset{\overset{\textstyle O}{\|}}{C} - C_{17}H_{35}
\end{array}
$$

| Tristearin | Glycerol | Sodium stearate (a soap) |

Because soaps are salts of strong bases and weak acids, they should be weakly alkaline in aqueous solution. However, a soap with free alkali can cause damage to skin, silk, or wool. Therefore, a test for basicity of the soap is quite important.

Soap has been largely replaced by synthetic detergents during the last two decades, because soap has two serious drawbacks. One is that soap becomes ineffective in hard water. Hard water contains appreciable amounts of Ca^{2+} or Mg^{2+} salts.

$$2C_{17}H_{35}COO^-Na^+ + M^{2+} \longrightarrow [C_{17}H_{35}COO^-]_2 \, M^{2+}\downarrow + 2Na^+$$

Soap   Scum

$$M = (Ca^{2+} \text{ or } Mg^{2+})$$

The other is that, in an acidic solution, soap is converted to free fatty acid and therefore loses its cleansing action.

$$C_{17}H_{35}COO^-Na^+ + H^+ \longrightarrow C_{17}H_{35}COOH\downarrow + Na^+$$

Soap   Fatty acid

Objectives

1. To prepare a simple soap.
2. To investigate some properties of a soap.

Procedure

Preparation of a Soap

Measure 23 mL of a vegetable oil into a 250-mL Erlenmeyer flask. Add 20 mL of ethyl alcohol (to act as a solvent) and 20 mL of 25% sodium hydroxide solution (25% NaOH). While stirring the mixture constantly with a glass rod, the flask with its contents is heated gently in a boiling water bath. A 600-mL beaker containing about 200 mL of tap water and a few boiling chips can serve as a water bath (Fig. 14.1).

Figure 14.1
Experimental set-up for
soap preparation.

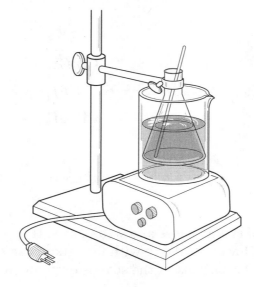

After being heated for about 20 min., the odor of alcohol will disappear, indicating the completion of the reaction. A pasty mass containing a mixture of the soap, glycerol, and excess sodium hydroxide is obtained. Use an ice water bath to cool the flask with its contents. To precipitate or "salt out" the soap, add 150 mL of a saturated sodium chloride solution to the soap mixture while stirring vigorously. This process increases the density of the aqueous solution; therefore, soap will float out from the aqueous solution. Filter the precipitated soap with the aid of suction and wash it with 10 mL of ice cold water. Observe the appearance of your soap and record your observation on the Report Sheet.

Properties of a Soap

1. *Emulsifying properties*. Shake 5 drops of mineral oil in a test tube containing 5 mL of water. A temporary emulsion of tiny oil droplets in water will be formed. Repeat the same test, but this time add a small piece of the soap you have prepared before shaking. Allow both solutions to stand for a short time. Compare the appearance and the relative stabilities of the two emulsions. Record your observations on the Report Sheet.

2. *Hard water reactions*. Place about one-third spatula full of the soap you have prepared in a 50-mL beaker containing 25 mL of water. Warm the beaker with its contents to dissolve the soap. Pour 5 mL of the soap solution into each of five test tubes (nos. 1, 2, 3, 4, and 5). Test no. 1 with 2 drops of a 5% solution of calcium chloride (5% $CaCl_2$), no. 2 with 2 drops of a 5% solution of magnesium chloride (5% $MgCl_2$), no. 3 with 2 drops of a 5% solution of iron(III) chloride (5% $FeCl_3$), and no. 4 with tap water. Tube no. 5 will be used for a basicity test, which will be performed later. Record your observations on the Report Sheet.

3. *Alkalinity (basicity)*. Test soap solution no. 5 with a wide-range pH paper. What is the approximate pH of your soap solution? Record your answer on the Report Sheet.

Chemicals and Equipment

1. Hot plate
2. Ice cubes
3. Büchner funnel in no. 7 one-hole rubber stopper
4. 500-mL filter flask
5. Filter paper, 7 cm diameter
6. pHydrion paper
7. Boiling chips
8. 95% ethanol
9. Saturated sodium chloride solution
10. 25% NaOH
11. Vegetable oil
12. 5% $FeCl_3$
13. 5% $CaCl_2$
14. Mineral oil
15. 5% $MgCl_2$

Experiment 14

PRE-LAB QUESTIONS

1. What chemical process is called saponification (soap making)? Why?

2. Consult Table 20.2 of your textbook. If corn oil is used to make soap, what is the chemical formula of the *most abundant* soap you formed?

3. How would you convert this soap to the corresponding fatty acid?

4. Stearic acid is insoluble in water, and sodium stearate (a soap) is soluble. What causes the difference in solubility? Explain.

Experiment 14

REPORT SHEET

Preparation

Appearance of your soap _____

Properties

Emulsifying Properties

Which mixture, oil–water or oil–water–soap, forms a more stable emulsion?

Hard Water Reaction

No. 1 + $CaCl_2$ _____

No. 2 + $MgCl_2$ _____

No. 3 + $FeCl_3$ _____

No. 4 + tap water _____

Alkalinity

pH of your soap solution (no. 5) _____

POST-LAB QUESTIONS

1. When you made soap, first you dissolved vegetable oil in ethanol. What happened to the ethanol during the reaction?

2. Write a chemical equation for the reaction in which you added a few drops of $MgCl_2$ solution to a soap solution.

3. Soaps that have a pH above 8.0 tend to irritate some sensitive skins. Was your soap good enough to compete with commercial preparations?

Preparation of a hand cream

Background

Hand creams are formulated to carry out a variety of cosmetic functions. Among these are softening the skin and preventing dryness; elimination of natural waste products (oils) by emulsification; cooling the skin by radiation thus helping to maintain body temperature. In addition, hand creams must have certain ingredients that aid spreadibility and provide body. In many cases added fragrance improves the odor, and in some special cases medications combat assorted ills.

The basic hand cream formulations all contain water to provide moisture and lanolin which helps its absorption by the skin. The latter is a yellowish wax. Chemically, wax is made of esters of long chain fatty acids and long chain alcohols. Lanolin is usually obtained from sheep wool; it has the ability to absorb 25–30% of its own weight of water and to form a fine emulsion. Mineral oil, which consists of high-molecular-weight hydrocarbons, provides spreadibility. In order to allow nonpolar substances, such as lanolin and mineral oil, to be uniformly dispersed in a polar medium, water, one needs strong emulsifying agents. An emulsifying agent must have nonpolar, hydrophobic portions to interact with the oil and also polar, hydrophilic portions to interact with water. A mixture of stearic acid and triethanolamine, through acid–base reaction, yields the salt that has the requirements to act as an emulsifying agent.

Besides the above five basic ingredients, some hand creams also contain alcohols such as propylene glycol (1,2-propanediol), and esters such as methyl stearate, to provide the desired texture of the hand cream.

In this experiment you will prepare four hand creams using the combination of ingredients as shown in Table 15.1.

Objectives

1. To learn the method of preparing a hand cream.
2. To appraise the function of the ingredients in the hand cream.

Procedure

Preparation of the Hand Creams

For each sample in Table 15.1, assemble the ingredients in two beakers. Beaker 1 contains the polar ingredients, and beaker 2 contains the nonpolar contents.

Table **15.1** **Recipes to Prepare Hand Creams**

Ingredients	Sample 1	Sample 2	Sample 3	Sample 4	
Water	25 mL	25 mL	25 mL	25 mL	
Triethanolamine	1 mL	1 mL	1 mL	—	Beaker 1
Propylene glycol	0.5 mL	0.5 mL	—	0.5 mL	
Stearic acid	5 g	5 g	5 g	5 g	
Methyl stearate	0.5 g	0.5 g	—	0.5 g	
Lanolin	4 g	4 g	4 g	4 g	Beaker 2
Mineral oil	5 mL	—	5 mL	5 mL	

1. To prepare sample 1, put the nonpolar ingredients in a 50-mL beaker (beaker 2) and heat it in a water bath. The water bath can be a 400-mL beaker half-filled with tap water and heated with a Bunsen burner (Fig. 15.1). Carefully hold the beaker with crucible tongs in the boiling water until all ingredients melt.

Figure 15.1
Heating ingredients.

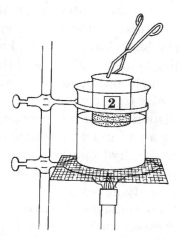

2. In the same water bath, heat the 100-mL beaker (beaker 1) containing the polar ingredients for about 5 min. Remove the beaker and set it on the bench top.

3. Into the 100-mL beaker containing polar ingredients, pour slowly the contents of the 50-mL beaker that holds the molten nonpolar ingredients (Fig. 15.2). Stir the mixture for 5 min. until you have a smooth uniform paste.

4. Repeat the same procedure in preparing the other three samples.

Figure 15.2
Mixing hand cream
ingredients.

Characterization of the Hand Cream Preparations

1. Test the pH of the hand creams prepared using a wide-range pH paper.

2. Rubbing a small amount of the hand cream between your fingers, test for smoothness and homogeneity. Also note the appearance. Record your observations on the Report Sheet.

3. Dispose of your hand cream preparations in the waste containers provided. **DO NOT** place in sink.

Chemicals and Equipment
1. Bunsen burner
2. Lanolin
3. Stearic acid
4. Methyl stearate
5. Mineral oil
6. Triethanolamine
7. Propylene glycol
8. pHydrion paper

Experiment 15

PRE-LAB QUESTIONS

1. Write the chemical formula of a wax made of an 18-carbon saturated fatty acid and a 10-carbon saturated straight chain alcohol.

2. The emulsifying agent was prepared from stearic acid and triethanolamine. Give the name of this salt. Write its formula.

3. What functional groups of the emulsifying agent provide the hydrophilic character?

4. What is the most abundant component of all hand creams?

NAME SECTION DATE

PARTNER GRADE

Experiment 15

REPORT SHEET

Characterization of the hand cream samples

Properties	Sample 1	Sample 2	Sample 3	Sample 4
pH				
Smoothness				
Homogeneity				
Appearance				

POST-LAB QUESTIONS

1. In comparing the properties of the hand creams you produced, ascertain the function of each of the missing ingredients in the hand cream:

(a) Mineral oil

(b) Triethanolamine

(c) Methyl stearate and propylene glycol

2. A hand cream appears smooth and uniform after you prepared it, but in a week of storage most of the water settles on the bottom and most of the oil separates on the top. What do you think may have gone wrong with the hand cream preparation?

3. Was the pH of all your hand cream preparation the same? If not, explain the differences.

4. In one of your hand cream formulation there was no mineral oil. What characteristics was observed in the absence of mineral oil? Explain.

Harcourt, Inc.

Extraction and identification
of fatty acids from corn oil

Background

Fats are esters of glycerol and fatty acids. Liquid fats are often called oils. Whether a fat is solid or liquid depends on the nature of the fatty acids. Solid animal fats contain mostly saturated fatty acids, while vegetable oils contain high amounts of unsaturated fatty acids. To avoid arteriosclerosis, hardening of the arteries, diets which are low in saturated fatty acids as well as in cholesterol are recommended.

Note that even solid fats contain some unsaturated fatty acids, and oils contain saturated fatty acids as well. Besides the degree of unsaturation, the length of the fatty acid chain also influences whether a fat is solid or liquid. Short chain fatty acids, such as found in coconut oil, convey liquid consistency in spite of the low unsaturated fatty acid content. Two of the unsaturated fatty acids, linoleic and linolenic acids, are essential fatty acids because the body cannot synthesize them from precursors; they must be included in the diet.

The four unsaturated fatty acids most frequently found in vegetable oils are:

Oleic acid: $CH_3(CH_2)_7CH = CH(CH_2)_7COOH$

Linoleic acid: $CH_3(CH_2)_4CH = CHCH_2CH = CH(CH_2)_7COOH$

Linolenic acid: $CH_3CH_2CH = CHCH_2CH = CHCH_2CH = CH(CH_2)_7COOH$

Arachidonic acid:
$CH_3(CH_2)_4CH = CHCH_2CH = CHCH_2CH = CHCH_2CH = CH(CH_2)_3COOH$

All the $C = C$ double bonds in the unsaturated fatty acids are *cis* double bonds, which interrupt the regular packing of the aliphatic chains, and thereby convey a liquid consistency at room temperature. This physical property of the unsaturated fatty acid is carried over to the physical properties of triglycerides (oils).

In order to extract and isolate fatty acids from corn oil, first, the ester linkages must be broken. This is achieved in the saponification reaction in which a triglyceride is converted to glycerol and the potassium salt of its fatty acids:

In order to separate the potassium salts of fatty acids from glycerol, the products of the saponification mixture must be acidified. Subsequently, the fatty acids can be extracted by petroleum ether. To identify the fatty acids that were isolated, they must be converted to their respective methyl ester by a perchloric acid catalyzed reaction:

$$C_{17}H_{35}COOH \ + \ CH_3OH \ \xrightarrow{HClO_4} \ C_{17}H_{35}\overset{\displaystyle O}{\overset{\|}{C}}{-}O{-}CH_3 \ + \ H_2O$$

The methyl esters of fatty acids can be separated by thin-layer chromatography (TLC). They can be identified by comparison of their rate of migration (R_f values) to the R_f values of authentic samples of methyl esters of different fatty acids (Fig. 16.1).

Figure 16.1
TLC chromatogram.

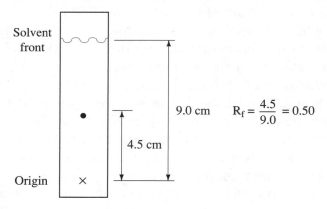

$R_f = $ distance travelled by fatty acid/distance travelled by the solvent front.

Objectives

1. To extract fatty acids from neutral fats.
2. To convert them to their methyl esters.
3. To identify them by thin-layer chromatography.

Procedure

Part A. Extraction of Fatty Acids

1. Weigh a 50-mL Erlenmeyer flask and record the weight on your Report Sheet (1).

2. Add 2 mL of corn oil and weigh it again. Record the weight on your Report Sheet (2).

3. Add 5 mL of 0.5 M KOH in ethanol to the Erlenmeyer flask. Stopper it. Place the flask in a water bath at 55°C for 20 min.

CAUTION!

Strong acid; use gloves with concentrated HCl.

---

4. When the saponification is completed, add 2.5 mL of the concentrated HCl. Mix it by swirling the Erlenmeyer flask. Transfer the contents into a 50-mL separatory funnel. Add 5 mL of petroleum ether. Mix it thoroughly (see Fig. 12.1). Drain the lower aqueous layer into a flask and the upper petroleum ether layer into a glass-stoppered test tube. Repeat the process by adding back the aqueous layer into the separatory funnel and extracting it with another portion of 5 mL of petroleum ether. Combine the extracts.

Part B. Preparation of Methyl Esters

1. Place a plug of glass wool (the size of a pea) into the upper stem of a funnel, fitting it loosely. Add 10 g of anhydrous Na_2SO_4. Rinse the salt on to the glass wool with 5 mL of petroleum ether; discard the wash. Pour the combined petroleum ether extracts into the funnel and collect the filtrate in an evaporating dish. Add another portion (2 mL) of petroleum ether to the funnel and collect this wash, also in the evaporating dish.

2. Evaporate the petroleum ether under the hood by placing the evaporating dish on a water bath at 60°C. (Alternatively, if dry N_2 gas is available, the evaporation could be achieved by bubbling nitrogen through the extract. This also must be done under the hood.)

3. When dry, add 10 mL of the $CH_3OH:HClO_4$ mixture (95:5). Place the evaporating dish in the water bath at 55°C for 10 min.

Part C. Identification of Fatty Acids

1. Transfer the methyl esters prepared above into a separatory funnel. Extract twice with 5 mL of petroleum ether. Combine the extracts.

2. Prepare another funnel with anhydrous Na_2SO_4 on top of the glass wool. Filter the combined petroleum ether extracts through the salt into a dry, clean evaporating dish. Evaporate the petroleum ether on the water bath at 60°C, as before. When dry, add 0.2 mL of petroleum ether and transfer the solution to a clean and dry test tube.

3. Take a 15 × 6.5 cm TLC plate. Make sure you do not touch the TLC plate with your fingers. Preferably use plastic gloves, or handle the plate by holding it only at the edges. This precaution must be observed throughout the whole operation because your fingers may contaminate the sample. With a pencil, lightly draw a line parallel to the 6.5 edge about 1 cm from the edge. Mark the positions of the five spots, equally spaced, where you will spot your samples (Fig. 16.2).

Figure 16.2
Spotting.

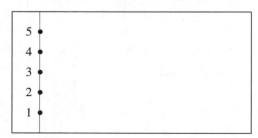

4. For spots no. 1 and no. 5, use your isolated methyl esters obtained from corn oil. For spot no. 2, use methyl oleate; for spot no. 3, methyl linoleate; and for spot no. 4, methyl palmitate. For each sample use a separate capillary tube. In spotting, apply each sample in the capillary to the plate until it spreads to a spot of 1 mm diameter. Dry the spots with a heat lamp. Pour about 15 mL of solvent (hexane:diethyl ether; 4:1) into a 500-mL beaker. Place the spotted TLC plate diagonally for ascending chromatography. Make certain that the spots applied are **above** the surface of the eluting solvent. Cover the beaker lightly with aluminum foil to avoid excessive solvent evaporation.

5. When the solvent front has risen to about 1–2 cm from the top edge, remove the plate from the beaker. Mark the advance of the solvent front with a pencil. Dry the plate with a heat lamp under the hood. Place the dried plate in a beaker containing a few iodine crystals. Cover the beaker tightly with aluminum foil. Place the beaker in a 110°C oven for 3–4 min. Remove the beaker and let it cool to room temperature. **This part is essential to avoid inhaling iodine vapors.** Remove the TLC plate from the beaker and mark the spots with a pencil.

6. Record the distance the solvent front advanced on your Report Sheet (4). Record on your Report Sheet (5–9) the distance of each iodine-stained spot from its origin. Calculate the R_f values of your samples (10–14).

Chemicals and Equipment

1. Corn oil
2. Methyl palmitate
3. Methyl oleate
4. Methyl linoleate
5. Petroleum ether (b. p. 30–60°C)
6. 0.5 M KOH in ethanol
7. Concentrated HCl
8. Anhydrous Na_2SO_4
9. Methanol:perchloric acid mixture (95:5)
10. Hexane:diethyl ether mixture (4:1)
11. Iodine crystals, I_2
12. Aluminum foil
13. Polyethylene gloves
14. 15 × 6.5 cm silica gel TLC plate
16. Capillary tubes open on both ends
17. Heat lamp
18. Water bath
19. Ruler
20. Drying oven, 110°C

Harcourt, Inc.

Experiment 16

PRE-LAB QUESTIONS

1. Fatty acids can be extracted by petroleum ether. Salts of fatty acids cannot; they are water soluble. Explain why.

2. Write the formulas of the reaction, converting linolenic acid to its methyl ester.

3. How can one convert the potassium salt of a fatty acid (i.e., potassium oleate) into a fatty acid (oleic acid)?

4. Why do you have to cool the iodine chamber (the beaker containing the chromatogram and iodine vapor) from 110°C to room temperature?

Experiment 16

REPORT SHEET

1. Weight of beaker _____ g
2. Weight of beaker and oil _____ g
3. Weight of oil _____ g

Distances on the chromatogram in cm

4. The solvent front _____
5. Spot no. 1 a, b, c, d, e   a____ b____ c____ d____ e____
6. Spot no. 2 _____
7. Spot no. 3 _____
8. Spot no. 4 _____
9. Spot no. 5 a, b, c, d, e   a____ b____ c____ d____ e____

Calculated R_f values

10. For spot no.1 [(5)/(4)] a, b, c, d, e   a____ b____ c____ d____ e____
11. For spot no. 2 [(6)/(4)] _____
12. For spot no. 3 [(7)/(4)] _____
13. For spot no. 4 [(8)/(4)] _____
14. For spot no. 5 [(9)/(4)] a, b, c, d, e   a____ b____ c____ d____ e____
15. How many fatty acids were present in your corn oil?

16. How many fatty acids could you identify? Name the identifiable fatty acids in the corn oil.

POST-LAB QUESTIONS

1. Which of the identifiable fatty acids of your corn oil was a saturated fatty acid?

2. Judging from the iodine spots of samples 2, 3, and 4, which fatty acid reacts most strongly with iodine? Why?

3. What was the role of the **anhydrous** Na_2SO_4 in the preparation of the methyl esters of fatty acids?

4. Given two saturated fatty acids, one a short chain of 10 carbons and the other a long chain of 20 carbons, which would move faster on the TLC plate? Explain.

5. Considering the R_f values you obtained for the three methyl esters of the fatty acids in your experiment, how could you achieve a better separation of the spots?

Experiment 17

Analysis of lipids

Background

Lipids are chemically heterogeneous mixtures. The only common property they have is their insolubility in water. We can test for the presence of various lipids by analyzing their chemical constituents. Foods contain a variety of lipids, most important among them are fats, complex lipids, and steroids. Fats are triglycerides, esters of fatty acids and glycerol. Complex lipids also contain fatty acids, but their alcohol may be either glycerol or sphingosine. They also contain other constituents such as phosphate, choline, or ethanolamine or mono- to oligo-saccharides. An important representative of this group is lecithin, a glycerophospholipid, containing fatty acids, glycerol, phosphate, and choline. The most important steroid in foods is cholesterol. Different foods contain different proportions of these three groups of lipids.

Structurally, cholesterol contains the steroid nucleus that is the common core of all steroids.

Steroid nucleus   Cholesterol

There is a special colorimetric test, the Lieberman-Burchard reaction, which uses acetic anhydride and sulfuric acid as reagents, that gives a characteristic green color in the presence of cholesterol. This color is due to the —OH group of cholesterol and the unsaturation found in the adjacent fused ring. The color change is gradual: first it appears as a pink coloration, changing later to lilac, and finally to deep green.

When lecithin is hydrolyzed in acidic medium, both the fatty acid ester bonds and the phosphate ester bonds are broken and free fatty acids and inorganic phosphate are released. Using a molybdate test, we can detect the presence of phosphate in the hydrolysate by the appearance of a purple color. Although this test is not specific for lecithin (other phosphate containing lipids will give a positive molybdate test), it differentiates clearly between fat and cholesterol on the one hand (negative test), and phospholipid on the other (positive test).

A second test that differentiates between cholesterol and lecithin is the acrolein reaction. When lipids containing glycerol are heated in the presence of potassium hydrogen sulfate, the glycerol is dehydrated, forming acrolein, which has an unpleasant odor. Further heating results in polymerization of acrolein, which is indicated by the slight blackening of the reaction mixture. Both the pungent smell and the black color indicate the presence of glycerol, and thereby fat and/or lecithin. Cholesterol gives a negative acrolein test.

$$
\begin{array}{c}
CH_2OH \\
| \\
CHOH \\
| \\
CH_2OH
\end{array}
\xrightarrow{\Delta}
\begin{array}{c}
O \\
\| \\
C-H \\
| \\
CH \\
\| \\
CH_2
\end{array}
+ 2H_2O
$$

Objectives

To investigate the lipid composition of common foods such as corn oil, butter, and egg yolk.

Procedure

Use six samples for each test: (1) pure cholesterol, (2) pure glycerol, (3) lecithin preparation, (4) corn oil, (5) butter, (6) egg yolk.

Phosphate Test

CAUTION!

6 M nitric acid is a strong acid. Handle it with care. Use gloves.

1. Take six clean and dry test tubes. Label them. Add about 0.2 g of sample to each test tube. Hydrolyze the compounds by adding 3 mL of 6 M nitric acid to each test tube.

2. Prepare a water bath by boiling about 100 mL of tap water in a 250-mL beaker on a hot plate. Place the test tubes in the boiling water bath for 5 min. Do not inhale the vapors. Cool the test tubes. Neutralize the acid by adding 3 mL of 6 M NaOH. Mix. During the hydrolysis, a precipitate may form, especially in the egg yolk sample. The samples in

which a precipitate appeared must be filtered. Place a piece of cheese cloth on top of a 25-mL Erlenmeyer flask. Pour the turbid hydrolysate in the test tube on the cheese cloth and filter it.

3. Transfer 2 mL of each neutralized (and filtered) sample into clean and labeled test tubes. Add 3 mL of a molybdate solution to each test tube and mix the contents. **(Be careful. The molybdate solution contains sulfuric acid.)** Heat the test tubes in a boiling water bath for 5 min. Cool them to room temperature.

4. Add 0.5 mL of an ascorbic acid solution and mix the contents thoroughly. Wait 20 min. for the development of the purple color. Record your observations on the Report Sheet. While you wait, you can perform the rest of the colorimetric tests.

The Acrolein Test for Glycerol

1. Place 1 g of potassium hydrogen sulfate, $KHSO_4$, in each of seven clean and dry test tubes. Label them. Add a few grains of your pure preparations, lecithin and cholesterol, to two of the test tubes. Add a drop, about 0.1 g, from each, glycerol, corn oil, butter, and egg yolk to the other four test tubes. To the seventh test tube add a few crystals of sucrose.

2. Set up your Bunsen burner in the hood. **It is important that this test be performed under the hood because of the pungent odor of the acrolein.**

3. Gently heat each test tube, one at a time, over the Bunsen burner flame, shaking it continuously from side to side. When the mixture melts it slightly blackens, and you will notice the evolution of fumes. Stop the heating. **Smell the test tubes by moving them sideways under your nose or waft the vapors. Do not inhale the fumes directly.** A pungent odor, resembling burnt hamburgers, is the positive test for glycerol. Sucrose in the seventh test tube also will be dehydrated and will give a black color. However, its smell is different, and thus is not a positive test for acrolein. Do not overheat the test tubes, for the residue will become hard, making it difficult to clean the test tubes. Record your observations on the Report Sheet.

Lieberman-Burchard Test for Cholesterol

1. Place a few grains of your cholesterol and lecithin preparations in labeled clean and dry test tubes. Similarly, add about 0.1-g samples of glycerol, corn oil, butter, and egg yolk to the other four labeled clean and dry test tubes. **(The next step should be done in the hood.)**

2. Transfer 3 mL of chloroform and 1 mL of acetic anhydride to each test tube. Finally, add 1 drop of concentrated sulfuric acid to each mixture. Mix the contents and record the color changes, if any. Wait 5 min. Record again the color of your solutions. Record your observations on the Report Sheet.

Chemicals and Equipment

1. 6 M NaOH
2. 6 M HNO$_3$
3. Molybdate reagent
4. Ascorbic acid solution
5. KHSO$_4$
6. Chloroform
7. Acetic anhydride
8. Sulfuric acid, H$_2$SO$_4$
9. Cholesterol
10. Lecithin
11. Glycerol
12. Corn oil
13. Butter
14. Egg yolk
15. Hot plate
16. Cheese cloth

Experiment 17

PRE-LAB QUESTIONS

1. Cephalins are glycerophospholipids present in foods. They differ from lecithins by having ethanolamine or serine instead of choline in their structure. Could you differentiate between lecithins and cephalins on the basis of the three tests to be performed in this experiment?

2. Cholesterol has an alcohol group. One could also dehydrate cholesterol (removing one water molecule by heating). Show the structure you would expect from the dehydration of cholesterol.

3. Would the compound with the structure in question 2 give a positive Lieberman-Burchard test?

4. Choleterol in tissues is sometimes esterified by fatty acids. (a) Draw the structure of cholesteryl oleate. (b) Would this ester give a positive Lieberman-Burchard test?

5. Why must you wear gloves in performing the phosphate test?

Experiment 17

REPORT SHEET

Tests	Cholesterol	Lecithin	Corn Glycerol	oil	Butter	Egg yolk	Sucrose
1. Phosphate **a.** Color							
b. Conclusions							
2. Acrolein **a.** Odor							
b. Color							
c. Conclusions							
3. Lieberman-Burchard **a.** Initial color							
b. Color after 5 min.							
c. Conclusion							

POST-LAB QUESTIONS

1. What is your overall conclusion regarding the composition of your corn oil? Was it pure triglyceride?

2. Based on the intensity of color developed in your test for cholesterol, which food contained the most and which contained the least cholesterol?

3. Besides the lecithin and other glycerophospholipids, two more classes of complex lipds are given in your textbook: (a) sphingolipids and (b) glycolipids. (Look up their structures in your textbook.) Would any of these compounds give you a positive test with molybdate solution?

4. A positive acrolein test is indicated by its odor as well as by its color. Which comes first? Explain.

5. When sucrose is dehydrated by heating it with $KHSO_4$, you can observe the black residue (carbon) and water. This is the origin of the name *carbohydrate*. Can you detect the presence of acrolein by its smell in the dehydration of sucrose?

6. Why was it necessary to hydrolyze the samples with nitric acid before performing the molybdate test?

Experiment 18

TLC separation of amino acids

Background

Amino acids are the building blocks of peptides and proteins. They possess two functional groups—the carboxylic acid group gives the acidic character, and the amino group provides the basic character. The common structure of all amino acids is

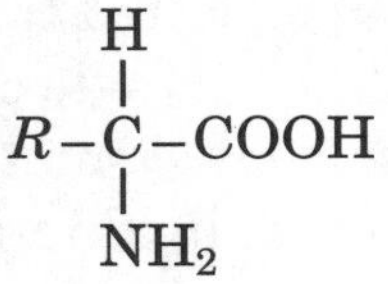

$$R-\overset{\overset{\displaystyle H}{|}}{\underset{\underset{\displaystyle NH_2}{|}}{C}}-COOH$$

The R represents the side chain that is different for each of the amino acids that are commonly found in proteins. However, all 20 amino acids have a free carboxylic acid group and a free amino (primary amine) group, except proline which has a cyclic side chain and a secondary amino group.

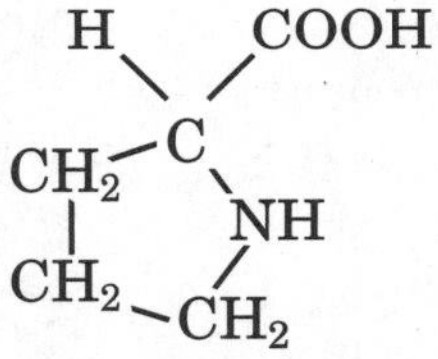

Proline

We use the properties provided by these groups to characterize the amino acids. The common carboxylic acid and amino groups provide the acid–base nature of the amino acids. The different side chains, and the solubilities provided by these side chains, can be utilized to identify the different amino acids by their rate of migration in thin-layer chromatography.

In this experiment, we use thin-layer chromatography to identify aspartame, an artificial sweetener, and its hydrolysis products from certain foods.

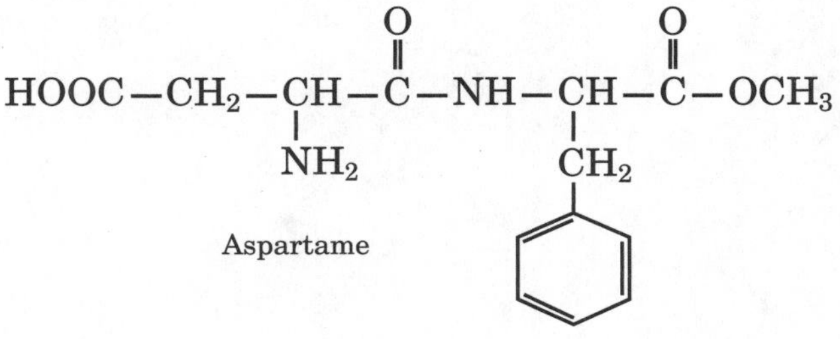

Aspartame

Aspartame is the methyl ester of the dipeptide aspartylphenylalanine. Upon hydrolysis with HCl it yields aspartic acid, phenylalanine, and methyl alcohol. When this artificial sweetener was approved by the Food and Drug Administration, opponents of aspartame claimed that it is a health hazard, because aspartame would be hydrolyzed and would yield poisonous methyl alcohol in soft drinks that are stored over long periods of time. The Food and Drug Administration ruled, however, that aspartame is sufficiently stable and fit for human consumption. Only a warning must be put on the labels of foods containing aspartame. This warning is for patients suffering from phenylketonurea who cannot tolerate phenylalanine.

To run a thin-layer chromatography experiment, we use silica gel in a thin layer on a plastic or glass plate. We apply the sample (aspartame or amino acids) as a spot to a strip of a thin-layer plate. The plate is dipped into a mixture of solvents. The solvent moves up the thin gel by capillary action and carries the sample with it. Each amino acid may have a different migration rate depending on the solubility of the side chain in the solvent. Amino acids with similar side chains are expected to move with similar, though not identical, rates; those that have quite different side chains are expected to migrate with different velocities. Depending on the solvent system used, almost all amino acids and dipeptides can be separated from each other by thin-layer chromatography (TLC).

We actually do not measure the rate of migration of an amino acid or a dipeptide, but rather, how far a particular amino acid travels in the thin silica gel layer relative to the migration of the solvent. This ratio is called the R_f value. In order to calculate the R_f values, one must be able to visualize the position of the amino acid or dipeptide. This is done by spraying the thin-layer silica gel plate with a ninhydrin solution that reacts with the amino group of the amino acid. A purple color is produced when the plate is heated. (The proline not having a primary amine gives a yellow color with ninhydrin.) For example, if the purple spot of an amino acid appears on the TLC plate 4.5 cm away from the origin and the solvent front migrates 9.0 cm (Fig. 18.1), the R_f value for the amino acid is calculated

$$R_f = \frac{\text{distance traveled by the amino acid}}{\text{distance traveled by the solvent front}} = \frac{4.5 \text{ cm}}{9.0 \text{ cm}} = 0.50$$

In the present experiment you will determine the R_f values of three amino acids: phenylalanine, aspartic acid, and leucine. You will also measure the R_f value of aspartame.

Figure 18.1
TLC chromatogram.

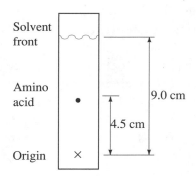

The aspartame you will analyze is actually a commercial sweetener, Equal by the NutraSweet Co., that contains silicon dioxide, glucose, cellulose, and calcium phosphate in addition to the aspartame. None of these other ingredients of Equal will give a purple or any other colored spot with ninhydrin. Other generic aspartame sweeteners may contain other nonsweetening ingredients. Occasionally, some sweeteners may contain a small amount of leucine which can be detected by the ninhydrin test. You will also hydrolyze aspartame using HCl as a catalyst to see if the hydrolysis products will prove that the sweetener is truly aspartame. Finally, you will analyze some commercial soft drinks supplied by your instructor. The analysis of the soft drink can tell you if the aspartame was hydrolyzed at all during the processing and storing of the soft drink.

Objectives

1. To separate amino acids and a dipeptide by TLC.
2. To identify hydrolysis products of aspartame.
3. To analyze the state of aspartame in soft drinks.

Procedure

1. Dissolve about 10 mg of the sweetener Equal in 1 mL of 3 M HCl in a test tube. Heat it with a Bunsen burner to a boil for 30 sec., but make sure that the liquid does not completely evaporate. Cool the test tube and label it "Hydrolyzed Aspartame."

2. Label five small test tubes, respectively, for aspartic acid, phenylalanine, leucine, aspartame, and Diet Coke. Place about 0.5-mL samples in each test tube.

3. Take two 15 × 6.5 cm TLC plates. With a pencil, lightly draw a line parallel to the 6.5 cm edge and about 1 cm from the edge. Mark the positions of five spots on each plate, spaced equally, where you will spot your samples (Fig. 18.2). **You must make sure that you don't touch the plates with your fingers.** Either use plastic gloves or handle the plates by holding them only at their edges. This precaution must be observed throughout the whole operation, because amino acids from your fingers will contaminate the plate.

Figure 18.2
Spotting.

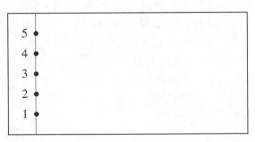

On plate A you will spot samples of (1) phenylalanine, (2) aspartic acid, (3) leucine, (4) aspartame in Equal, and (5) the hydrolyzed aspartame you prepared in step no. 1. On plate B you will spot samples of Diet Coke on lanes (1) and (4), aspartic acid on lane (2), aspartame in Equal on lane (3), and the hydrolyzed aspartame you prepared previously on lane (5).

4. First spot plate A. For each sample use a separate capillary tube. Apply the sample to the plate until it spreads to a spot of 1 mm diameter. Dry the spots. (If a heat lamp is available, use it for drying.) Pour about 15 mL of solvent mixture (butanol:acetic acid:water) into a large (500-mL or 1-L) beaker and place your spotted plate diagonally for an ascending chromatography. Make certain that the spots applied to the plate are above the surface of the eluting solvent. Cover the beaker with aluminum foil to avoid the evaporation of the solvent mixture.

5. Spot plate B. For aspartic acid, lane (2), and for the hydrolyzed and nonhydrolyzed aspartame, lanes (3) and (5), use one spot as before. For Diet Coke [lanes (1) and (4)] multiple spotting is needed. Apply the capillary tube 12–15 times to the same spot, making certain that between each application the previous sample has been dried. Also, try to control the size of the spots so that they do not spread too much, not more than 2 mm in diameter. Dry the spots as before. Place the plate in a large beaker containing the eluting solvent as before. Cover the beaker with aluminum foil. Allow about 50–60 min. for the solvent front to advance.

6. When the solvent front nears the edge of the plate, about 1–2 cm from the edge, remove the plate from the beaker. You must not allow the solvent front to advance up to or beyond the edge of the plate. Mark immediately *with a pencil* the position of the solvent front. Under a hood dry the plates with the aid of a heat lamp or hair dryer. Using polyethylene gloves, spray the dry plates with ninhydrin solution. *Be careful not to spray ninhydrin on your hand and not to touch the sprayed areas with bare hands. If the ninhydrin spray touches your skin (which contains amino acids) your fingers will be discolored for a few days.* Place the sprayed plates into a drying oven at 105–110°C for 2–3 min.

7. Remove the plates from the oven. Mark the center of the spots and calculate the R_f values of each spot. Record your observations on the Report Sheet.

8. If the spots on the chromatogram are faded, we can visualize them by exposing the chromatogram to iodine vapor. Place your chromatogram into a wide-mouthed jar containing a few iodine crystals. Cap the jar and warm it slightly on a hot plate to enhance the sublimation of iodine. The iodine vapor will interact with the faded pigment spots and make them visible. After a few minutes' exposure to iodine vapor, remove the chromatogram and mark the spots **immediately** with a pencil. The spots will fade again with exposure to air. Measure the distance of the center of the spots from the origin and calculate the R_f values.

CAUTION!

For The Instructor: With some batches of TLC plates the solvent front may move too slowly. As an alternative, chromatography paper (Whatman chromatography paper no. 1, 0.016 mm thickness) can be substituted. In this case the solvent front should not be allowed to move farther than 60 mm from the origin. The spotted chromatography paper should be taped with Scotch tape to a glass rod and suspended into the eluting solvent. Be certain that the liquid level is below the spots applied to the paper. The remaining steps are the same.

Chemicals and Equipment

1. 0.1% solutions of aspartic acid, phenylalanine, and leucine
2. 0.5% solution of aspartame (Equal)
3. Diet Coke
4. 3 M HCl
5. 0.2% ninhydrin spray
6. Butanol:acetic acid:water–solvent mixture
7. Equal sweetener
8. Aluminum foil
9. 15 × 6.5 cm silica gel TLC plates
10. Ruler
11. Polyethylene gloves
12. Capillary tubes open on both ends
13. Heat lamp or hair dryer
14. Drying oven, 110°C
15. Wide-mouthed jar
16. Iodine crystals

Experiment 18

PRE-LAB QUESTIONS

1. If an amino acid has an R_f value of 0.45, how far will the amino acid move on a TLC plate in which the solvent front moved 15.2 cm?

2. All amino acids give a purple color when stained with ninhydrin. Only proline gives a yellowish color. Can you give a reason why this amino acid stains differently?

3. What happens if you don't use gloves and your finger comes in contact with the ninhydrin spray?

4. Write the structure of the mono- and dimethyl ester of aspartic acid.

Experiment 18

REPORT SHEET

1.

Plate A	Distance traveled (mm)	Solvent front (mm)	R_f
Phenylalanine			
Aspartic acid			
Leucine			
Aspartame			
Hydrolyzed aspartame			

Plate B	Distance traveled (mm)	Solvent front (mm)	R_f
Diet Coke			
Aspartic acid			
Aspartame			
Diet Coke			
Hydrolyzed aspartame			

2. Identification

(a) Name the amino acids you found in the hydrolysate of the sweetener Equal.

(b) How many spots were stained with ninhydrin (1) in Equal and (2) in Diet Coke samples?

POST-LAB QUESTIONS

1. Your laboratory period had only 90 min. for the development of the chromatogram. In order to get better separation of the spots you must allow the solvent front to move much farther than the value you reported on your Report Sheet. Assuming a steady rate of solvent movement, how long of a lab period do you need for the solvent front to move 12.5 cm?

2. In testing the hydrolysate of aspartame, you forgot to mark the position of the solvent front on your TLC plate. Could you

   (a) determine how many amino acids were in the aspartame;

   (b) identify those amino acids?

3. Do you have any evidence that the aspartame was hydrolyzed during the processing and storage of the Diet Coke sample? Explain.

4. The difference between aspartic acid and phenylalanine is twofold. Aspartic acid has a polar, acidic side chain, while phenylalanine has a nonpolar side chain. The molecular weight of aspartic acid is smaller than the molecular weight of phenylalanine. Based on the R_f values you obtained for these two amino acids in the solvent employed, which property influenced the rate of migration?

5. The R_f value of leucine is somewhat smaller than that of phenylalanine. Both are nonpolar amino acids. Leucine has a smaller molecular mass than alanine so you would expect it to move faster. Yet it is moving slower. How could you explain your results?

Experiment 19

Acid–base properties of amino acids

Background

In the body, amino acids exist as zwitterions.

$$R-\overset{\overset{\displaystyle H}{|}}{\underset{\underset{\displaystyle NH_3^+}{|}}{C}}-COO^-$$

This is an amphoteric compound because it behaves as both an acid and a base in the Brønsted definition. As an acid, it can donate an H^+ and becomes the conjugate base:

$$R-\overset{\overset{\displaystyle H}{|}}{\underset{\underset{\displaystyle NH_3^+}{|}}{C}}-COO^- + OH^- \rightleftharpoons R-\overset{\overset{\displaystyle H}{|}}{\underset{\underset{\displaystyle NH_2}{|}}{C}}-COO^- + H_2O$$

| Acid | Base | Conj. base | Conj. acid |

As a base, it can accept an H^+ ion and becomes the conjugate acid:

$$R-\overset{\overset{\displaystyle H}{|}}{\underset{\underset{\displaystyle NH_3^+}{|}}{C}}-COO^- + H_3O^+ \rightleftharpoons R-\overset{\overset{\displaystyle H}{|}}{\underset{\underset{\displaystyle NH_3^+}{|}}{C}}-COOH + H_2O$$

| Base | Acid | Conj. acid | Conj. base |

To study the acid–base properties, one can perform a simple titration. We start our titration with the amino acid being in its acidic form at a low pH:

$$R-\overset{\overset{\displaystyle H}{|}}{\underset{\underset{\displaystyle NH_3^+}{|}}{C}}-COOH \quad (I)$$

As we add a base, OH^-, to the solution, the pH will rise. We record the pH of the solution by using a pH meter after each addition of the base. To obtain the titration curve, we plot the milliliters of NaOH added against the pH of the solution (Fig. 19.1).

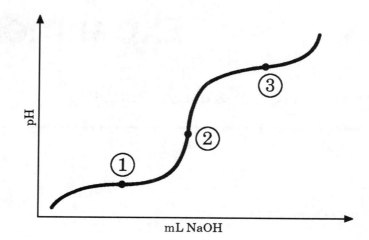

Figure 19.1
The titration curve
of an amino acid.

Note that there are two flat portions (called legs) on the titration curve where the pH does not increase appreciably with the addition of NaOH. The midpoint of the first leg, ①, is when half of the original acidic amino acid (I) has been titrated and it becomes a zwitterion (II).

$$\begin{array}{c} H \\ | \\ R-C-COO^- \quad (II) \\ | \\ NH_3{}^+ \end{array}$$

The point of inflection, ②, occurs when the amino acid is entirely in the zwitterion form (II). At the midpoint of the second leg, ③, half of the amino acid is in the zwitterion form and half is in the basic form (III).

$$\begin{array}{c} H \\ | \\ R-C-COO^- \quad (III) \\ | \\ NH_2 \end{array}$$

From the pH at the midpoint of the first leg we obtain the pK value of the carboxylic acid group, since this is the group that is titrated with NaOH at this stage (the structure going from I to II). The pH of the midpoint of the second leg, ③, is equal to the pK of the $-NH_3{}^+$, since this is the functional group that donates its H^+ at this stage of the titration. The pH at the inflection point, ②, is equal to the isoelectric point. At the isoelectric point of a compound, the positive and negative charges balance each other. This occurs at the inflection point when all the amino acids are in the zwitterion form.

You will obtain a titration curve of an amino acid with a neutral side chain such as glycine, alanine, phenylalanine, leucine, or valine. If pH meters are available, you read the pH directly from the instrument after each addition of the base. If a pH meter is not available, you can obtain the pH with the aid of indicator papers. From the titration curve obtained, you can determine the pK values and the isoelectric point.

1. To study acid–base properties by titration.
2. To calculate pK values for the titratable groups.

Procedure

1. Pipet 20 mL of 0.1 M amino acid solution (glycine, alanine, phenylalanine, leucine, or valine) that has been acidified with HCl to a pH of 1.5 into a 100-mL beaker.

2. If a pH meter is available, insert the clean and dry electrode of the pH meter into a standard buffer solution with known pH. Turn the knob of the meter to the pH mark and adjust it to read the pH of the buffer. Turn the knob of the pH meter to "Standby" position. Remove the electrode from the buffer, wash it with distilled water, and dry it. Insert the dry electrode into the amino acid solution. Turn the knob of the meter to "pH" position and record the pH of the solution. Fill a buret with 0.25 M NaOH solution. Add the NaOH solution from the buret in 1.0-mL increments to the beaker. After each increment, stir the contents with a glass rod and then read the pH of the solution. Record these on your Report Sheet. Continue the titration as described until you reach pH 12. Turn off your pH meter, wash the electrode with distilled water, wipe it dry, and store it in its original buffer.

3. If a pH meter is not available, perform the titration as above, but use pH indicator papers. After the addition of each increment and stirring, withdraw a drop of the solution with a Pasteur pipet. Touch the end of the pipet to a dry piece of the pH indicator paper. Compare the color of the indicator paper with the color on the charts supplied. Read the corresponding pH from the chart and record it on your Report Sheet.

4. Draw your titration curve. From the graph, determine your pK values and the isoelectric point of your amino acid. Record these on your Report Sheet.

Chemicals and Equipment

1. 0.1 M amino acid solution (glycine, alanine, leucine, phenylalanine, or valine)
2. 0.25 M NaOH solution
3. pH meter and standard buffer (or pH indicator paper and Pasteur pipet)
4. 50-mL buret
5. 20-mL pipet
6. Spectroline pipet filler

Experiment 19

PRE-LAB QUESTIONS

1. In titrating the acidic form of an amino acid with NaOH solution, at which point in the titration curve does it become a zwitterion?

2. If the equilibrium constant, K_a, for the ionization of the carboxylic acid group is 1×10^{-3}, what is the pK_a?

3. If in a solution of alanine, the number of negatively charged carboxylate groups, $-COO^-$, is 1×10^{20} at the isoelectric point, what is the number of positively charged amino groups, NH_3^+?

Experiment 19

REPORT SHEET

1. Amino acid used for titration _____

mL of 0.25 M NaOH added	pH	mL of 0.25 M NaOH added	pH
0		13.0	
1.0		14.0	
2.0		15.0	
3.0		16.0	
4.0		17.0	
5.0		18.0	
6.0		19.0	
7.0		20.0	
8.0		21.0	
9.0		22.0	
10.0		23.00	
11.0		24.0	
12.0		25.00	

2. Plot your data below to get the titration curve.

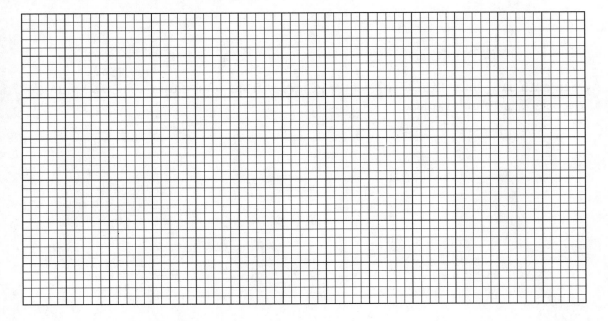

3. a. Indicate the positions of the midpoints of each leg and the position of the inflection point on your graph.

  b. Record the pK values for the carboxylic acid group _____, and for the amino group _____.

  c. Record the pH of the isoelectric point _____.

POST-LAB QUESTIONS

1. The isoelectric point of an amino acid is an intensive property.

  (a) Knowing that, would you expect to find your inflection point at a different pH value, if you had titrated 0.5 M solution of the same amino acid instead of the 0.1 M solution? Explain.

  (b) Would your result be different if you had used 50 mL of amino acid solution instead of 20 mL? Explain.

2. Check the pI values of the different amino acids in your textbook (Table 21.1). On the basis of the isoelectric point (pI) obtained in your experiment, how would you classify the amino acid of your experiment?

3. Which data can you obtain with greater accuracy from your graph—the pK values or the isoelectric point? Explain.

Isolation and identification of casein

Background

Casein is the most important protein in milk. It functions as a storage protein, fulfilling nutritional requirements. Casein can be isolated from milk by acidification to bring it to its isoelectric point. At the isoelectric point, the number of positive charges on a protein equals the number of negative charges. Proteins are least soluble in water at their isoelectric points because they tend to aggregate by electrostatic interaction. The positive end of one protein molecule attracts the negative end of another protein molecule, and the aggregates precipitate out of solution.

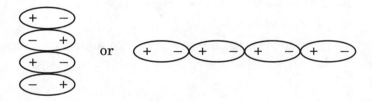

On the other hand, if a protein molecule has a net positive charge (at low pH or acidic condition) or a net negative charge (at high pH or basic condition), its solubility in water is increased.

$$\overset{+}{N}H_3 \sim COOH \underset{\text{low pH}}{\overset{H^+}{\longleftarrow}} \overset{+}{N}H_3 \sim COO^- \underset{\text{high pH}}{\overset{OH^-}{\longrightarrow}} NH_2 \sim COO^- + H_2O$$

$\quad$ More soluble $\qquad\qquad$ Least soluble $\qquad\qquad$ More soluble
$\qquad\qquad\qquad\qquad$ (at isolelectric pH)

In the first part of this experiment, you are going to isolate casein from milk which has a pH of about 7. Casein will be separated as an insoluble precipitate by acidification of the milk to its isoelectric point (pH = 4.6). The fat that precipitates along with casein can be removed by dissolving it in alcohol.

In the second part of this experiment, you are going to prove that the precipitated milk product is a protein. The identification will be achieved by performing a few important chemical tests.

1. *The biuret test.* This is one of the most general tests for proteins. When a protein reacts with copper(II) sulfate, a positive test is the formation of a copper complex which has a violet color.

Protein   Blue color   Protein–copper complex
(violet color)

This test works for any protein or compound that contains two or more of the following groups:

$$-\overset{\overset{\displaystyle O}{\|}}{C}-NH-, \quad -\overset{\overset{\displaystyle O}{\|}}{C}-NH_2, \quad -CH_2-NH_2, \quad -\overset{\overset{\displaystyle NH}{\|}}{C}-NH_2, \quad -\overset{\overset{\displaystyle S}{\|}}{C}-NH_2$$

2. *The ninhydrin test.* Amino acids with a free $-NH_2$ group and proteins containing free amino groups react with ninhydrin to give a purple-blue complex.

Amino acid   Ninhydrin

Purple-blue complex

3. *Heavy metal ions test.* Heavy metal ions precipitate proteins from solution. The ions that are most commonly used for protein precipitation are Zn^{2+}, Fe^{3+}, Cu^{2+}, Sb^{3+}, Ag^+, Cd^{2+}, and Pb^{2+}. Among these metal ions, Hg^{2+}, Cd^{2+}, and Pb^{2+} are known for their notorious toxicity to humans. They can cause serious damage to proteins (especially

enzymes) by denaturing them. This can result in death. The precipitation occurs because proteins become cross-linked by heavy metals as shown below:

$$2NH_2 \text{~~~~} \overset{\overset{\displaystyle O}{\parallel}}{C}-O^- + Hg^{2+} \longrightarrow \left\{ \begin{array}{c} H_2N \\ \\ C-O \\ \parallel \\ O \end{array} \quad Hg \quad \begin{array}{c} O-C \\ \parallel \\ O \\ \\ NH_2 \end{array} \right\}$$

Insoluble precipitate

Victims swallowing Hg^{2+} or Pb^{2+} ions are often treated with an antidote of a food rich in proteins, which can combine with mercury or lead ions in the victim's stomach and, hopefully, prevent absorption! Milk and raw egg white are used most often. The insoluble complexes are then immediately removed from the stomach by an emetic.

4. *The xanthoprotein test.* This is a characteristic reaction of proteins that contain phenyl rings

Concentrated nitric acid reacts with the phenyl ring to give a yellow-colored aromatic nitro compound. Addition of alkali at this point will deepen the color to orange.

$$HO-\langle\bigcirc\rangle-CH_2-\underset{\underset{\displaystyle H}{\vert}}{\overset{\overset{\displaystyle NH_2}{\vert}}{C}}-COOH + HNO_3 \longrightarrow HO-\langle\bigcirc\rangle\overset{NO_2}{}-CH_2-\underset{\underset{\displaystyle H}{\vert}}{\overset{\overset{\displaystyle NH_2}{\vert}}{C}}-COOH + H_2O$$

Tyrosine   Colored compound

The yellow stains on the skin caused by nitric acid are the result of the xanthoprotein reaction.

Objectives

1. To isolate the casein from milk under isoelectric conditions.
2. To perform some chemical tests to identify proteins.

Procedure

Part A: Isolation of Casein

1. To a 250-mL Erlenmeyer flask, add 50.00 g of milk and heat the flask in a water bath (a 600-mL beaker containing about 200 mL of tap water; see Fig. 20.1). Stir the solution

constantly with a stirring rod. When the bath temperature has reached about 40°C, remove the flask from the water bath, and add about 10 drops of glacial acetic acid while stirring. Observe the formation of a precipitate.

Figure 20.1
Precipitation of casein.

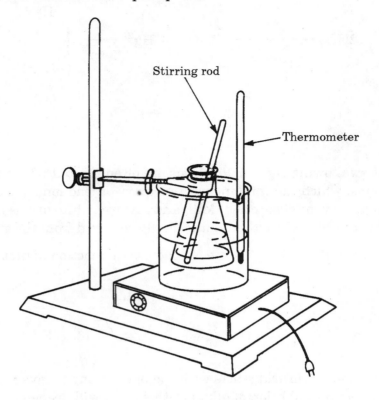

2. Filter the mixture into a 100-mL beaker by pouring it through a cheese cloth which is fastened with a rubber band over the mouth of the beaker (Fig. 20.2). Remove most of the water from the precipitate by squeezing the cloth gently. Discard the filtrate in the beaker. Using a spatula, scrape the precipitate from the cheese cloth into the empty flask.

Figure 20.2
Filtration of casein.

3. Add 25 mL of 95% ethanol to the flask. After stirring the mixture for 5 min., allow the solid to settle. Carefully decant (pour off) the liquid that contains fats into a beaker. Discard the liquid.

4. To the residue, add 25 mL of a 1:1 mixture of diethyl ether-ethanol. After stirring the resulting mixture for 5 min., collect the solid by vacuum filtration.

CAUTION:

Diethyl ether is highly flammable. Make sure there is no open flame in the lab.

---

5. Spread the casein on a paper towel and let it dry. Weigh the dried casein and calculate the percentage of casein in the milk. Record it on your Report Sheet.

$$\% \text{ casein} = \frac{\text{weight of solid (casein)}}{50.00 \text{ g of milk}} \times 100$$

Part B: Chemical Analysis of Proteins

1. *The biuret test.* Place 15 drops of each of the following solutions in five clean, labeled test tubes.

a. 2% glycine

b. 2% gelatin

c. 2% albumin

d. Casein prepared in Part A (one-quarter of a full spatula) + 15 drops of distilled water

e. 1% tyrosine

To each of the test tubes, add 5 drops of 10% NaOH solution and 2 drops of a dilute $CuSO_4$ solution while swirling. The development of a purplish-violet color is evidence of the presence of proteins. Record your results on the Report Sheet.

2. *The ninhydrin test.* Place 15 drops of each of the following solutions in five clean, labeled test tubes.

a. 2% glycine

b. 2% gelatin

c. 2% albumin

d. Casein prepared in Part A (one-quarter of a full spatula) + 15 drops of distilled water

e. 1% tyrosine

To each of the test tubes, add 5 drops of ninhydrin reagent and heat the test tubes in a boiling water bath for about 5 min. Record your results on the Report Sheet.

3. *Heavy metal ions test.* Place 2 mL of milk in each of three clean, labeled test tubes. Add a few drops of each of the following metal ions to the corresponding test tubes as indicated below:

a. Pb^{2+} as $Pb(NO_3)_2$ in test tube no. 1

b. Hg^{2+} as $Hg(NO_3)_2$ in test tube no. 2

c. Na^+ as $NaNO_3$ in test tube no. 3

Record your results on the Report Sheet.

CAUTION!

The following test will be performed by your instructor.

4. *The xanthoprotein test.* (Perform the experiment under the hood.) Place 15 drops of each of the following solutions in five clean, labeled test tubes:

   a. 2% glycine

   b. 2% gelatin

   c. 2% albumin

   d. Casein prepared in Part A (one-quarter of a full spatula) + 15 drops of distilled water

   e. 1% tyrosine

   To each test tube, add 10 drops of concentrated HNO_3 while swirling. Heat the test tubes carefully in a warm water bath. Observe any change in color. Record the results on your Report Sheet.

Chemicals and Equipment

1. Hot plate
2. Büchner funnel in a no. 7 one-hole rubber stopper
3. 500-mL filter flask
4. Filter paper (Whatman no. 2, 7 cm)
5. Cheese cloth
6. Rubber band
7. Boiling chips
8. 95% ethanol
9. Diethyl ether–ethanol mixture
10. Regular milk
11. Glacial acetic acid
12. Concentrated nitric acid
13. 2% albumin
14. 2% gelatin
15. 2% glycine
16. 5% copper(II) sulfate
17. 5% lead(II) nitrate
18. 5% mercury(II) nitrate
19. Ninhydrin reagent
20. 10% sodium hydroxide
21. 1% tyrosine
22. 5% sodium nitrate

Experiment 20

PRE-LAB QUESTIONS

1. Casein has an isoelectric point at pH 4.6. What kind of charges will be on the casein in its native environment, that is, in milk?

2. How do you separate the fat from the protein in the casein precipitate?

3. Would the amino acid, glycine, give a positive biuret test? Explain.

4. What are the three most toxic heavy metal ions?

Experiment 20

REPORT SHEET

Isolation of casein

1. Weight of milk   _____ g
2. Weight of dried casein   _____ g
3. Percentage of casein in milk   _____ %

Chemical analysis of proteins

Biuret test

Substance	Color formed
2% glycine	
2% gelatin	
2% albumin	
casein + H_2O	
1% tyrosine	

Which of these chemicals gives a positive test with this reagent?   _____

Ninhydrin test

Substance	Color formed after heating
2% glycine	
2% gelatin	
2% albumin	
casein + H_2O	
1% tyrosine	

Which of these chemicals gives a positive test with this reagent?   _____

Heavy metal ion test

Substance	Precipitates formed
$Pb(NO_3)_2$	
$Hg(NO_3)_2$	
$NaNO_3$	

Which of these metal ions gives a positive test with casein in milk? _____

Xanthoprotein test

Substance	Color formed before or after heating
2% glycine	
2% gelatin	
2% albumin	
casein + H_2O	
1% tyrosine	

Which of these chemicals gives a positive test with this reagent? _____

POST-LAB QUESTIONS

1. Explain why casein precipitates when acetic acid is added to it.

2. In the isolation of casein following the acidification, you removed the precipitate by filtering through a cheese cloth and squeezing the cloth. If you did not squeeze out all the liquids, would your yield of casein be different? Explain.

3. Does gelatin contain tyrosine? Explain.

4. If by mistake (*don't try it*) your finger touches nitric acid and you observe a yellow color on your fingers, what functional group(s) in your skin is (are) responsible for this reaction?

5. Why is milk or raw egg used as an antidote in cases of heavy metal ion poisoning?

6. According to your results, how many grams of casein are in a glass of milk (175 g)?

Isolation and identification of DNA from yeast

Background

Hereditary traits are transmitted by genes. Genes are parts of giant deoxyribonucleic acid (DNA) molecules. In lower organisms, such as bacteria and yeast, both DNA and RNA (ribonucleic acid) occur in the cytoplasm. In higher organisms, most of the DNA is inside the nucleus, and the RNA is outside the nucleus in other organelles and in the cytoplasm.

In this experiment, we will isolate DNA molecules from yeast cells. The first task is to break up the cells. This is achieved by a combination of different techniques and agents. Grinding up the cells with sand disrupts them and the cytoplasm of many yeast cells is spilled out. However, this is not a complete process. The addition of a detergent, hexadecyltrimethylammonium bromide, CTAB, accomplishes two functions: (1) it helps to solubilize cell membranes and thereby further weakens the cell structure, and (2) it helps to inactivate the nucleic acid-degrading enzymes, nucleases, that are present. The addition of a chelating agent, ethylenediamine tetraacetate, EDTA, also inactivates these enzymes. EDTA removes the di- and tri-valent cations necessary for the activity of nucleases. Without this inhibition, the nucleases would degrade the nucleic acids to their constituent nucleotides. The final assault on the yeast cell is the osmotic shock. This is provided by a hypotonic saline–EDTA solution. The already weakened cells (by grinding and treatment with CTAB) will burst in the hypotonic medium and spill their contents, nucleic acids, among them.

Once the nucleic acids are in solution, they must be separated from the other constituents of the cell. First, the protein molecules must be removed. Many of the proteins of the cell are strongly associated with nucleic acids. The addition of sodium perchlorate ($NaClO_4$) dissociates the proteins from nucleic acids. When the mixture is shaken with the organic solvent, chloroform-isoamyl alcohol, the proteins are denatured, and they precipitate at the interface. At the same time, the lipid components of the cells are dissolved in the organic solvent. Thus the aqueous layer will contain nucleic acids, small water-soluble molecules, and even some proteins as contaminants.

The addition of ethanol precipitates the large molecules (DNA, RNA, and proteins) and leaves the small molecules in solution. DNA, being the largest fibrous molecule, forms thread-like precipitates that can be spooled off onto a rod. The protein and RNA form a gelatinous precipitate that cannot be picked up by winding them on a glass rod. Thus, the spooling separates DNA from RNA.

After the isolation of DNA, we will probe its identity by using the diphenylamine test. The blue color of this test is specific for deoxyribose and the appearance of a blue color can be used to identify the deoxyribose-containing DNA molecule.

Flow Diagram of the DNA Isolation Process

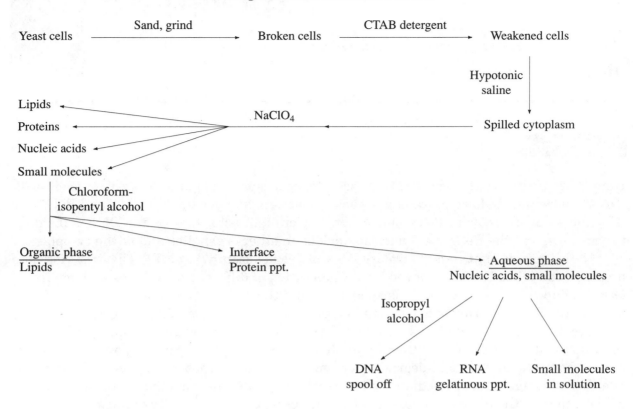

Objectives

To demonstrate the separation of DNA molecules from other cell constituents and to prove their identity.

Procedure

1. Cool a mortar in ice water. Add 2 to 3 g of baker's yeast and twice as much acid-washed sand. Grind the yeast and the sand vigorously with a pestle for 5–10 min. to disrupt the cells. (Two groups can work together in grinding; then divide the product.)

2. Preheat 25 mL of hexadecyltrimethylammonium bromide (CTAB) isolation buffer (2% CTAB, 0.15 M NaCl, 0.2% 2-mercaptoethanol, 20 mM EDTA, and 100 mM Tris-HCl at pH 8.0) in a 100-mL beaker in a 60°C water bath.

3. Add the ground yeast and sand to the saline–CTAB solution. Mix the solution with the sand. Let it stand for 20 min., with occasional swirling, while maintaining the temperature at 60°C.

4. Decant the cell suspension into a 250-mL Erlenmeyer flask, leaving the sand behind. Cool the solution to room temperature. Add 5 mL of 6 M $NaClO_4$ solution and mix well. Transfer 40 mL of the chloroform-isopentyl alcohol mixture into the flask. Stopper the flask with a cork. Shake it for 10 min., sloshing the contents from side to side once every 15 sec. A frothy emulsion will form. After 10 min., let the emulsion settle.

5. Break up the emulsion by gently swirling with a glass rod that reaches into the interface. The complete separation into two distinct layers is not possible without centrifugation. (If desk top centrifuges are available, it is preferable to separate the layers by centrifuging at 1600× gravity for 5 min.) However, one can proceed without centrifugation as well. When a sufficient amount (20–30 mL) of the top aqueous layer is cleared, remove this with a Pasteur pipet and transfer it to a graduated cylinder. Measure the volume and pour the contents into a 250-mL beaker. Pay attention that none of the brownish precipitate, droplets of emulsion, is transferred.

6. To the viscous DNA-containing aqueous solution, add slowly twice its volume of cold isopropyl alcohol, taking care that the alcohol flows along the side of the beaker, settling on top of the aqueous solution. With a flame-sterilized glass rod, gently stir the DNA-isopropyl alcohol solution. This procedure is **critical**. The DNA will form a thread-like precipitate. **Rotating (not stirring) the glass rod** spools all the DNA precipitate onto the glass rod. As the DNA is wound on the rod, squeeze out the excess liquid by pressing the rod against the wall of the beaker. Transfer the spooled DNA on the rod into a test tube containing 95% ethanol.

7. Discard the alcohol solution left in the beaker and the chloroform–isoamyl alcohol solution left in the Erlenmeyer flask into specially labeled waste jars. Do not pour them down the sink.

8. Remove the rod and the spooled DNA from the test tube. Dry the DNA with a clean filter paper. Note its appearance. Dissolve the isolated crude DNA in 2 mL of citrate buffer (0.15 M NaCl, 0.015 M sodium citrate). Set up four dry and clean test tubes. Add 2 mL each of the following to the test tubes:

Test tube	Solution
1	1% glucose
2	1% ribose
3	1% deoxyribose
4	crude DNA solution

CAUTION!

Diphenylamine reagent contains glacial acetic acid and concentrated sulfuric acid. Handle with care. Use gloves.

Add 5 mL diphenylamine reagent to each test tube. Mix the contents of the test tubes. Heat the test tubes in boiling water bath for 10 min. Record the color on your Report Sheet.

Chemicals and Equipment

1. Baker's yeast
2. Sand
3. Saline-hexadecyltrimethylammonium bromide (CTAB) isolation buffer
4. $NaClO_4$ solution
5. Chloroform-isopentyl alcohol solvent
6. Citrate buffer
7. Isopropyl alcohol (2-propanol)
8. Glucose solution
9. Ribose solution
10. Deoxyribose solution
11. Diphenylamine reagent
12. 95% ethanol
13. Mortar and pestle
14. Desk top clinical centrifuges (optional)

Experiment 21

PRE-LAB QUESTIONS

Consult your textbook to answer the structural questions.

1. Draw the structures of adenine and thymine. Show the hydrogen bonds that may hold together this base pair.

2. DNA is strongly associated with proteins (especially histones). How can one remove the proteins to isolate pure DNA?

3. Why must you handle the diphenylamine reagent with great care?

4. The most demanding part of this experiment is grinding the yeast cells with sand. What does this process accomplish? Would you be able to isolate DNA without this grinding?

NAME ______________________ SECTION ____________ DATE ____________

PARTNER ______________________ GRADE ______________________

Experiment 21

REPORT SHEET

1. Describe the appearance of the crude DNA preparation.

2. Diphenylamine test.

Solution	**Color**
1% glucose	______________________
1% ribose	______________________
1% deoxyribose	______________________
crude DNA sample	______________________

Did the diphenylamine test confirm the identity of DNA?

3. Did you obtain a thread-like precipitate of DNA with isopropyl alcohol? Were the threads long enough so that you were able to spool the DNA onto the glass rod?

4. After mixing the aqueous extract with chloroform–isoamyl alcohol mixture, which layer contained the RNA (aqueous or organic)?

5. What compounds were left behind in the isopropyl alcohol solution after spooling the DNA?

POST-LAB QUESTIONS

1. Can the diphenylamine reagent distinguish between ribose and deoxyribose, and between DNA and RNA?

2. Write the structure of CTAB. Do not look it up in handbooks, only consider the full name of the compound. This compound acted as a detergent in your isolation procedure. Label which part of the CTAB structure is polar and which part is nonpolar.

3. Why can we isolate DNA from the precipitate, which also contains RNA and proteins, by the simple "spooling" procedure?

Viscosity and secondary structrue of DNA

Background

In 1953, Watson and Crick proposed a three-dimensional structure of DNA which is a cornerstone in the history of biochemistry and molecular biology. The double helix they proposed for the secondary structure of DNA gained immediate acceptance, partly because it explained all known facts about DNA, and partly because it provided a beautiful model for DNA replication.

In the DNA double helix, two polynucleotide chains run in opposite directions. This means that at each end of the double helix there is one 5′-OH and one 3′-OH terminal. The sugar phosphate backbone is on the outside, and the bases point inward. These bases are paired so that for each adenine (A) on one chain a thymine (T) is aligned opposite it on the other chain. Each cytosine (C) on one chain has a guanine (G) aligned with it on the other chain. The AT and GC base pairs form hydrogen bonds with each other. The AT pair has two hydrogen bonds; the GC pair has three hydrogen bonds (Fig. 22.1).

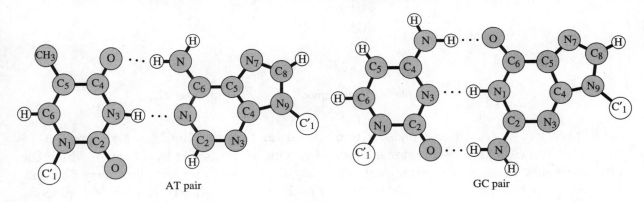

AT pair   GC pair

Figure 22.1 • Hydrogen bonding between base pairs.

Most of the DNA in nature has the double helical secondary structure. The hydrogen bonds between the base pairs provide the stability of the double helix. Under certain conditions the hydrogen bonds are broken. During the replication process itself, this happens and parts of the double helix unfold. Under other conditions, the whole molecule unfolds, becomes single stranded, and assumes a random coil conformation. This can happen in denaturation processes aided by heat, extreme acidic or basic conditions, etc. Such a transformation is often referred to as helix-to-coil transition. There are a number of techniques that can monitor such a transition. One of the most sensitive is the measurement of viscosity of DNA solutions.

Viscosity is the resistance to flow of a liquid. Honey has a high viscosity and gasoline a low viscosity, at room temperature. In a liquid flow, the molecules must slide past each other. The resistance to flow comes from the interaction between the molecules as they slide past each other. The stronger this interaction, i.e., hydrogen bonds vs. London dispersion forces, the greater the resistance and the higher the viscosity. Even more than the nature of the intermolecular interaction, the size and the shape of the molecules influence their viscosity. A large molecule has greater surface over which it interacts with other molecules than a small molecule. Therefore, its viscosity is greater than that of a small molecule. If two molecules have the same size and the same interaction forces but have different shapes, their viscosity will be different. For example, needle-shaped molecules, when aligned parallel by the flow of liquid, have greater surfaces of interaction than spherical molecules of the same molecular weight (Fig. 22.2). The needle-shaped molecule will have a higher viscosity than the spherical molecule. The DNA double helix is a rigid structure held together by hydrogen bonds. Its long axis along the helix exceeds by far its short axis perpendicular to it. Thus the DNA double helix has large surface area and consequently high viscosity. When the hydrogen bonds are broken and the DNA molecule becomes single stranded, it assumes a random coil shape which has much lower surface area and lower viscosity. Thus a helix-to-coil transition is accompanied by a drop in viscosity.

Figure 22.2
Surface area of interaction between molecules of different shapes.

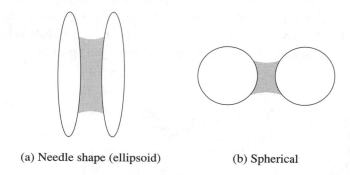

(a) Needle shape (ellipsoid)   (b) Spherical

In practice, we can measure viscosity by the efflux time of a liquid in a viscometer (Fig. 22.3). The capillary viscometer is made of two bulbs connected by a tube in which the liquid must flow through a capillary tube. The capillary tube provides a laminary flow in which concentric layers of the liquid slide past each other. Originally, the liquid is placed in the storage bulb (A). By applying suction above the capillary, the liquid is sucked up past the upper calibration mark. With a stopwatch in hand, the suction is released and the liquid is allowed to flow under the force of gravity. The timing starts when the meniscus of the liquid hits the upper calibration mark. The timing ends when the meniscus of the liquid hits the lower calibration mark of the viscometer. The time elapsed between these two marks is the efflux time.

Harcourt, Inc.

Figure 22.3
Ostwald capillary viscometer.

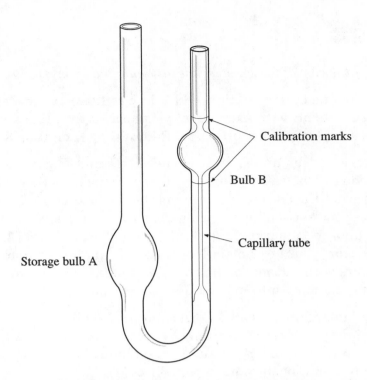

With dilute solutions, such as the DNA in this experiment, the viscosity of the solution is compared to the viscosity of the solvent. The efflux time of the solvent, aqueous buffer, is t_o and that of the solution is t_s. The relative viscosity of the solution is

$$\eta_{rel} = t_s/t_o$$

The viscosity of a solution also depends on the concentration; the higher the concentration, the higher the viscosity. In order to make the measurement independent of concentration, a new viscometric parameter is used which is called intrinsic viscosity, $[\eta]$. This number is

$$[\eta] = (\log \eta_{rel})/c$$

which is almost a constant for a particular solute (DNA in our case) in very dilute solutions.

In this experiment, we follow the change in the viscosity of a DNA solution when we change the pH of the solution from the very acidic (pH 2.0) to very basic (pH 12.0). At extreme pH values, we expect that the hydrogen bonds will break and the double helix will become single-stranded random coils. A change in the viscosity will tell at what pH this happens. We shall also determine whether two acid-denatured single-stranded DNA molecules can refold themselves into a double helix when we neutralize the denaturing acid.

Objectives

1. To demonstrate helix-to-coil-to-helix transitions.
2. To learn how to measure viscosity.

Procedure

Because of the cost of viscometers the students may work in groups of 5–6.

1. To 3 mL of a buffer solution, add 1 drop of 1.0 M HCl using a Pasteur pipet. Measure its pH with a universal pH paper. If the pH is above 2.5, add another drop of 1 M HCl. Measure the pH again. Record the pH on your Report Sheet (1).

2. Clamp one clean and dry viscometer on a stand. Pipet 3 mL of your acidified buffer solution into bulb A of your viscometer. Using a suction bulb of a Spectroline pipet filler, raise the level of the liquid in the viscometer above the upper calibration mark. Release the suction by removing the suction bulb and time the efflux time between the two calibration marks. Record this as t_o on your Report Sheet (2). Remove all the liquid from your viscometer by pouring the liquid out from the wide arm. Then apply pressure with the suction bulb on the capillary arm of the viscometer and blow out any remaining liquid into the storage bulb (A); pour out this residual liquid.

3. Take 3 mL of the prepared DNA solution. Add the same amount of 1 M HCl as above (1 or 2 drops). Mix it thoroughly by shaking the solution. Test the pH of the solution with a universal pH paper and record the pH (3) and the DNA concentration of the prepared solution on your Report Sheet (4).

4. Pour the acidified DNA solution into the wide arm (bulb A) of your viscometer. Using a suction bulb, raise the level of your liquid above the upper calibration mark. Release the suction by removing the suction bulb and measure and record the efflux time of the acidified DNA solution (5).

5. Add the same amount (1 or 2 drops) as above of neutralizing 1 M NaOH solution to the liquid in the wide arm of your viscometer. With the suction bulb on the capillary arm blow a few air bubbles through the solution to mix the ingredients. Repeat the measurement of the efflux time and record it on your Report Sheet (6). For the next 100 min. or so, repeat the measurement of the efflux times every 20 min. and record the results on your Report Sheet (7–11).

6. While the efflux time measurements in viscometer no. 1 are repeated every 20 min., another dry and clean viscometer will be used for establishing the pH dependence of the viscosity of DNA solutions. First, measure the pH of the buffer solution with a universal pH paper. Record it on your Report Sheet (12). Second, transfer 3 mL of the buffer into the viscometer no. 2 and measure and record its efflux time (13). Empty the viscometer as instructed in no. 2 above. Test the pH of the DNA solution with a universal pH paper (14) and transfer 3 mL into the viscometer. Measure its efflux time and record it on your Report Sheet (15). Empty your viscometer.

7. Repeat the procedure described in step no. 6, but this time, with the aid of a Pasteur pipet, add one drop of 0.1 M HCl both to the 3-mL buffer solution, as well as to the 3-mL DNA solution. Measure the pH and the efflux times of both buffer and DNA solutions and record them (16–19) on your Report Sheet. *Make sure that you empty the viscometer after each viscosity measurement.*

8. Repeat the procedure described in step no. 6, but this time add one drop of 0.1 M NaOH solution to both the 3-mL buffer and 3-mL DNA solutions. Measure their pH and efflux times and record them on your Report Sheet (20–23).

9. Repeat the procedure described in step no. 6, but this time add 2 drops of 1 M NaOH to both buffer and DNA solutions (3 mL of each solution). Measure and record their pH and efflux times on your Report Sheet (24–27).

10. If time allows, you may repeat the procedure at other pH values; for example, by adding two drops of 1 M HCl (28–31), or two drops of 0.1 M HCl (32–35), or two drops of 0.1 M NaOH (36–39) to the separate samples of buffer and DNA solutions.

Chemicals and Equipments

1. Viscometers, 3-mL capacity
2. Stopwatch or watch with a second hand
3. Stand with utility clamp
4. Pasteur pipets
5. Buffer at pH 7.0
6. Prepared DNA solution
7. 1 M HCl
8. 0.1 M HCl
9. 1 M NaOH
10. 0.1 M NaOH
11. Spectroline pipet fillers

Experiment 22

PRE-LAB QUESTIONS

1. Write an equation for the reaction between an amine, $-NH_2$, and an acid, H_3O^+.

2. Show the GC base pair structure before and after the addition of an acid.

3. Write an equation for the reaction between an ammonium cation, $-NH_3^+$, and a base, $-OH^-$.

4. Show the GC base pair structure after addition of a base, OH^-, to the acidified DNA.

5. Which will have a greater surface of interaction—DNA in a double helix or the same DNA denatured, single-stranded random coil? Justify your answer with a diagram of helix-to-coil transition.

Experiment 22

REPORT SHEET

1. pH of acidified buffer   _____

2. Efflux time of acidified buffer   _____ sec.

3. pH of acidified DNA solution   _____

4. Concentration of DNA solution   _____

5. Efflux time of acidified DNA solution   _____ sec.

6. Efflux time of neutralized DNA solution
   at time of neutralization   _____ sec.

7. 20 min. later   _____ sec.

8. 40 min. later   _____ sec.

9. 60 min. later   _____ sec.

10. 80 min. later   _____ sec.

11. 100 min. later   _____ sec.

12. pH of neutral buffer   _____

13. Efflux time of neutral buffer   _____ sec.

14. pH of DNA solution in neutral buffer   _____

15. Efflux time of DNA in neutral buffer   _____ sec.

After addition of 1 drop of 0.1 M HCl

16. pH of buffer   _____

17. Efflux time of buffer   _____ sec.

18. pH of DNA solution   _____

19. Efflux time of DNA solution   _____ sec.

After addition of 1 drop of 0.1 M NaOH

20. pH of buffer _____

21. Efflux time of buffer _____ sec.

22. pH of DNA solution _____

23. Efflux time of DNA solution _____ sec.

After addition of 2 drops of 1 M NaOH

24. pH of buffer _____

25. Efflux time of buffer _____ sec.

26. pH of DNA solution _____

27. Efflux time of DNA solution _____ sec.

After addition of 2 drops of 1 M HCl

28. pH of buffer _____

29. Efflux time of buffer _____ sec.

30. pH of DNA solution _____

31. Efflux time of DNA solution _____ sec.

After addition of 2 drops of 0.1 M HCl

32. pH of buffer _____

33. Efflux time of buffer _____ sec.

34. pH of DNA solution _____

35. Efflux time of DNA solution _____ sec.

After addition of 2 drops of 0.1 M NaOH

36. pH of buffer _____

37. Efflux time of buffer _____ sec.

38. pH of DNA solution _____

39. Efflux time of DNA solution _____ sec.

Tabulate your data on the pH dependence of relative viscosity.

pH	η_{rel}
(3) _____	(5)/(2) _____
(14) _____	(15)/(13) _____
(18) _____	(19)/(17) _____
(22) _____	(23)/(21) _____
(26) _____	(27)/(25) _____
(30) _____	(31)/(29) _____
(34) _____	(35)/(33) _____
(38) _____	(39)/(37) _____

POST-LAB QUESTIONS

1. Plot your tabulated data—relative viscosity on the y-axis and pH on the x-axis.

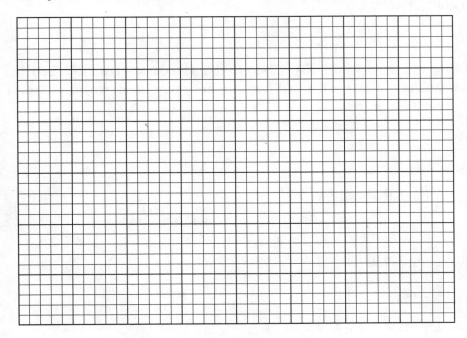

2. At what pH values did you observe helix-to-coil transitions?

3. Plot your data on the refolding of DNA double helix (5)–(11). Plot the time on the x-axis and the efflux times on the y-axis.

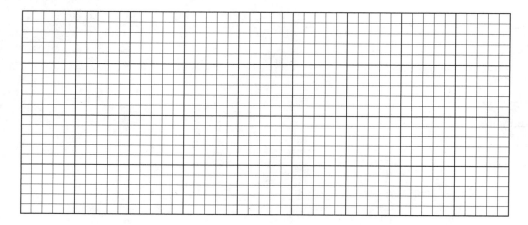

4. Was there any indication that, upon neutralization of the denaturing acid, the DNA did refold into a double helix? Explain.

5. Compare the efflux time of the neutral DNA (15) to that of the denatured DNA 100 min. after neutralization (11). What does the difference between these two efflux times tell you regarding the refolding process?

6. Calculate the intrinsic viscosity of your DNA at

   (a) neutral pH = 2.3 × {log[(15)/(13)]}/(4) =

   (b) acidic pH = 2.3 × {log[(5)/(2)]}/(4) =

   (c) basic pH = 2.3 × {log[(27)/(25)]}/(4) =

   (d) neutralized pH 100 min. after neutralization = 2.3 × {log[(11)/(13)]}/(4) =

7. A high intrinsic viscosity implies a double helix, a low intrinsic viscosity means a random coil. What do you think is the shape of the DNA after acid denaturation and subsequent neutralization? [See 6(d).] Explain your answer.

Kinetics of urease—
catalyzed decomposition of urea

Background

Enzymes speed up the rates of reactions by forming an enzyme-substrate complex. The reactants can undergo the reaction on the surface of the enzyme, rather than finding each other by collision. Thus the enzyme lowers the energy of activation of the reaction.

Urea decomposes according to the following equation:

$$\underset{\text{H}_2\text{N}-\overset{\displaystyle \text{O}}{\overset{\|}{\text{C}}}-\text{NH}_2}{} + \text{H}_2\text{O} \rightleftharpoons \text{CO}_2 + 2\text{NH}_3 \qquad (1)$$

This reaction is catalyzed by a highly specific enzyme, urease. Urease is present in a number of bacteria and plants. The most common source of the enzyme is jack bean or soybean. Urease was the first enzyme that was crystallized. Sumner, in 1926, proved unequivocally that enzymes are protein molecules.

Urease is an -SH group (thiol) containing enzyme. The cysteine residues of the protein molecule must be in the reduced -SH form in order for the enzyme to be active. Oxidation of these groups will form -S-S-, disulfide bridges, and the enzyme loses its activity. Reducing agents such as cysteine or glutathione can reactivate the enzyme.

Heavy metals such as Ag^+, Hg^{2+}, or Pb^{2+}, which form complexes with the -SH groups, also inactivate the enzyme. For example, the poison phenylmercuric acetate is a potent inhibitor of urease.

$$\text{enzyme}-\text{SH} + \text{CH}_3-\overset{\displaystyle \text{O}}{\overset{\|}{\text{C}}}-\text{O}^-\text{Hg}^+\text{C}_6\text{H}_5 \rightleftharpoons \text{CH}_3\overset{\displaystyle \text{O}}{\overset{\|}{\text{C}}}-\text{OH} + \text{enzyme}-\text{S}-\text{Hg}-\text{C}_6\text{H}_5 \qquad (2)$$

| Active | Phenylmercuric acetate | Acetic acid | Inactive |

In this experiment, we study the kinetics of the urea decomposition. As shown in equation (1), the products of the reaction are carbon dioxide, CO_2, and ammonia, NH_3. Ammonia, being a base, can be titrated with an acid, HCl, and in this way we can determine the amount of NH_3 that is produced.

$$\text{NH}_3(\text{aq}) + \text{HCl}(\text{aq}) \rightleftharpoons \text{NH}_4\text{Cl}(\text{aq}) \qquad (3)$$

For example, a 5-mL aliquot of the reaction mixture is taken before the reaction starts. We use this as a blank. We titrate this with 0.05 N HCl to an end point. The amount of acid used was 1.5 mL. This blank then must be subtracted from all subsequent titration values. Next, we take a 5-mL sample of the reaction mixture after the reaction has proceeded for 10 min. We titrate this with 0.05 N HCl and, let's assume, get a value of 5.0 mL HCl. Therefore, $5.0 - 1.5 = 3.5$ mL of 0.05 N HCl was used to neutralize the NH_3 produced in a 10-min. reaction time. This means that

$$(3.5 \text{ mL} \times 0.05 \text{ moles HCl})/1000 \text{ mL} = 1.75 \times 10^{-4} \text{ moles HCl}$$

was used up. According to reaction (3), one mole of HCl neutralizes 1 mole of NH_3, therefore, the titration indicates that in our 5-mL sample, 1.75×10^{-4} moles of NH_3 was produced in 10 min. Equation (1) also shows that for each mole of urea decomposed, 2 moles of NH_3 are formed. Therefore, in 10 min.

$$(1 \text{ mole urea} \times 1.75 \times 10^{-4} \text{ moles } NH_3)/2 \text{ moles } NH_3$$
$$= 0.87 \times 10^{-4} \text{ moles urea or } 8.7 \times 10^{-5} \text{ moles of urea}$$

were decomposed. Thus the rate was 8.7×10^{-6} moles of urea per min. This is the result we obtained using a 5-mL sample in which 1 mg of urease was dissolved. This rate of reaction corresponds to 8.7×10^{-6} moles urea/mg enzyme-min.

A *unit of activity* of urease is defined as the micromoles (1×10^{-6} moles) of urea decomposed in 1 min. Thus the enzyme in the preceding example had an activity of 8.7 units per mg enzyme.

In this experiment we also study the rate of the urease-catalyzed decomposition in the presence of an inhibitor. We use a dilute solution of phenylmercuric acetate to inhibit but not completely inactivate urease.

CAUTION!

Mercury compounds are poisons. Take extra care to avoid getting the mercuric salt solution in your mouth or swallowing it.

Many of the enzymes in our body are also -SH-containing enzymes, and these will be inactivated if we ingest such compounds. As a result of mercury poisoning, many body functions will be inhibited.

Objectives

1. To demonstrate how to measure the rate of an enzyme-catalyzed reaction.
2. To investigate the effect of an inhibitor on the rate of reaction.
3. To calculate urease activity.

Procedure

Enzyme Kinetics in the Absence of Inhibitor

1. Prepare a 37°C water bath in a 250-mL beaker. Maintain this temperature by occasionally adding hot water to the bath. To a 100-mL Erlenmeyer flask, add 20 mL of

0.05 M Tris buffer and 20 mL of 0.3 M urea in a Tris buffer. Mix the two solutions, and place the corked Erlenmeyer flask into the water bath for 5 min. This is your *reaction vessel*.

2. Set up a buret filled with 0.05 N HCl. Place into a 100-mL Erlenmeyer flask 3 to 4 drops of a 1% $HgCl_2$ solution. This will serve to stop the reaction, once the sample is pipetted into the titration flask. Add a few drops of methyl red indicator. This Erlenmeyer flask will be referred to as the *titration vessel*.

3. Take the *reaction vessel* from the water bath. Add 10 mL of urease solution to your reaction vessel. The urease solution contains a specified amount of enzyme (e.g., 20 mg enzyme in 10 mL of solution). Note the time of adding the enzyme solution as zero reaction time. Immediately pipet a 5-mL aliquot of the urea mixture into your *titration vessel*. Stopper the reaction vessel, and put it back into the 37°C bath.

4. Titrate the contents of the titration vessel with 0.05 N HCl to an end point. The end point is reached when the color changes from yellow to pink and stays that way for 10 sec. Record the amount of acid used. This is your blank.

5. Wash and rinse your titration vessel after each titration and reuse it for subsequent titrations.

6. Take a 5-mL aliquot from the reaction vessel **every 10 min**. Pipet these aliquots into the cleaned titration vessel into which methyl red indicator and $HgCl_2$ inhibitor were already placed similar to the procedure in step no. 2 that you used in your first titration (blank). Record the time you placed the aliquots into the titration vessels and titrate them with HCl to an end point. Record the amount of HCl used in your titration. Use five samples over a period of 50 min.

Enzyme Kinetics in the Presence of Inhibitor

CAUTION!

Be careful with the phenylmercuric acetate solution. Do not get it in your mouth or eyes.

---

1. Use the same water bath as in the first experiment. Maintain the temperature at 37°C. To a new 100-mL reaction vessel, add 19 mL of 0.05 M Tris buffer, 20 mL of 0.3 M urea solution, and 1 mL of phenylmercuric acetate (1×10^{-3} M). Mix the contents, and place the reaction vessel into the water bath for 5 min.

2. Ready the *titration vessel* as before by adding a few drops of $HgCl_2$ and methyl red indicator. To the *reaction vessel*, add 10 mL of urease solution. Note the time of addition as zero reaction time. Mix the contents of the *reaction vessel*. Transfer immediately a 5-mL aliquot into the *titration vessel*. This will serve as your blank.

3. Titrate it as before. Record the result. **Every 10 min.** take a 5-mL aliquot for titration. The duration of this experiment should be 40 min.

Chemicals and Equipment

1. Tris buffer
2. 0.3 M urea
3. 0.05 N HCl
4. 1×10^{-3} M phenylmercuric acetate
5. 1% $HgCl_2$
6. Methyl red indicator
7. Urease solution
8. 50-mL buret
9. 10-mL graduated pipets
10. 5-mL volumetric pipets
11. 10-mL volumetric pipets
12. Buret holder
13. Spectroline pipet filler

Experiment 23

PRE-LAB QUESTIONS

1. Consider the classification of enzymes in your textbook (Sect. 22.1) and reaction (1) of this experiment. How would you classify urease?

2. The decomposition of urea yields to gases, but you will not see gas bubbles forming in the reaction. Why is that so?

3. Why is phenylmercuric acetate such a dangerous poison?

4. How do we measure the concentration of a product from the urease-catalyzed reaction?

Experiment 23

REPORT SHEET

Enzyme kinetics in the absence of inhibitor

Reaction time (min.)	Buret readings before titration (A)	Buret readings after titration (B)	mL acid titrated (B) − (A)	mL 0.05 N HCl used up in the reaction (B) − (A) − blank
0 (blank)				
10				
20				
30				
40				
50				

Enzyme kinetics in the presence of Hg salt inhibitor

Reaction time (min.)	Buret readings before titration (A)	Buret readings after titration (B)	mL acid titrated (B) − (A)	mL 0.05 N HCl used up in the reaction (B) − (A) − blank
0 (blank)				
10				
20				
30				
40				

1. Present the preceding data in the graphical form by plotting reaction time (column 1) on the x-axis and the mL of 0.5 N HCl used (column 5) on the y-axis for both reactions.

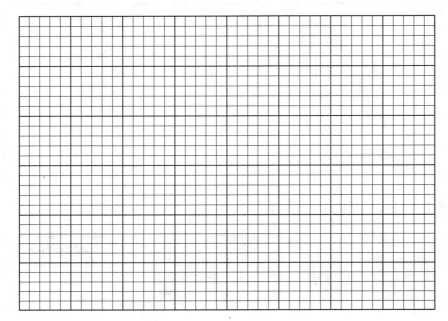

2. Calculate the urease activity only for the reaction without the inhibitor. Use the titration data from the first 10 min. of reaction (initial slope).

Urease activity

$$= \frac{\text{X mL HCl consumed} \times 0.05 \text{ moles HCl} \times 1 \text{ mole NH}_3 \times 1 \text{ mole urea} \times 50 \text{ mL sol.}}{10 \text{ min.} \times 5 \text{ mL sol.} \times 1000 \text{ mL HCl} \times 1 \text{ mole HCl} \times 2 \text{ moles NH}_3 \times 20 \text{ mg urease}}$$

$= Z$ units activity/mg enzyme

POST-LAB QUESTIONS

1. What would be the urease activity if you used the slope between 40 and 50 min. instead of the initial slope from your diagram?

2. Your instructor will provide the activity of urease as it was specified by the manufacturer. Compare this activity with the one you calculated. Can you account for the difference? (Enzymes usually lose their activity in long storage.)

3. If you want to perform the urease inhibition experiment but you cannot obtain $HgCl_2$, what could you substitute as an enzyme inhibitor?

4. Is the heavy metal inhibition of urease a reversible or irreversible inhibition? Explain.

5. In studying enzyme reactions, you must work at constant temperature and pH. What steps were taken in your experiment to satisfy these requirements?

6. Your lab ran out of 0.05 N HCl. You found a bottle labeled 0.05 N H_2SO_4. Could you use it for your titration? If you do, would your calculation of urease activity be different?

7. Compare the initial rates (the first 10 min.) of the enzyme reactions with and without inhibitor. How many times slower was the reaction with inhibitor than without it?

Isocitrate dehydrogenase—an enzyme of the citric acid cycle

Background

The citric acid cycle is the first unit of the common metabolic pathway through which most of our food is oxidized to yield energy. In the citric acid cycle, the partially fragmented food products are broken down further. The carbons of the C_2 fragments are oxidized to CO_2, released as such, and expelled in the respiration. The hydrogens and the electrons of the C_2 fragments are transferred to the coenzyme, nicotinamide adenine dinucleotide, NAD^+, or to flavin adenine dinucleotide, FAD, which in turn become $NADH + H^+$ or $FADH_2$, respectively. These enter the second part of the common pathway, oxidative phosphorylation, and yield water and energy in the form of ATP.

The first enzyme of the citric acid cycle to catalyze both the release of one carbon dioxide and the reduction of NAD^+ is isocitrate dehydrogenase. The overall reaction of this step is as follows:

$$
\begin{array}{c}
COO^- \\
| \\
CH_2 \\
| \\
CH-COO^- \\
| \\
HO-CH \\
| \\
COO^-
\end{array}
+ NAD^+ \xrightarrow{\text{enzyme}}
\begin{array}{c}
COO^- \\
| \\
CH_2 \\
| \\
CH_2 \\
| \\
C=O \\
| \\
COO^-
\end{array}
+ NADH + CO_2
$$

Isocitrate α-Ketoglutarate

The reduction of the NAD^+ itself is given by the equation:

$$
\text{NAD}^+ + H^+ + 2\,e^- \rightleftharpoons \text{NADH}
$$

NAD$^+$ NADH

The enzyme has been isolated from many tissues, the best source being a heart muscle or yeast. The isocitrate dehydrogenase requires the presence of cofactors Mg^{2+} or Mn^{2+}. As an allosteric enzyme, it is regulated by a number of modulators. ADP, adenosine diphosphate, is a positive modulator and therefore stimulates enzyme activity. The enzyme has an optimum pH of 7.0. As is the case with all enzymes of the citric acid cycle, isocitrate dehydrogenase is found in the mitochondria.

In the present experiment, you will determine the activity of isocitrate dehydrogenase extracted from pork heart muscle. The commercial preparation comes in powder form and it uses $NADP^+$ rather than NAD^+ as a coenzyme. The basis of the measurement of the enzyme activity is the absorption spectrum of NADPH. This reduced coenzyme has an absorption maximum at 340 nm. Therefore, an increase in the absorbance at 340 nm indicates an increase in NADPH concentration, hence the progress of the reaction. We define the unit of isocitrate dehydrogenase activity as one that causes an increase of 0.01 absorbance per min. at 340 nm.

For example, if a 10-mL solution containing isocitrate and isocitrate dehydrogenase and $NADP^+$ exhibits a 0.04 change in the absorbance in 2 min., the enzyme activity will be

$$\frac{0.04 \text{ abs.}}{2 \text{ min.} \times 10 \text{ mL}} \times \frac{1 \text{ unit}}{0.01 \text{ abs./1 min.}} = 0.2 \text{ units/mL}$$

If the 10-mL test solution contained 1 mL of isocitrate dehydrogenase solution with a concentration of 1 mg powder/1 mL of enzyme solution, then the activity will be

(0.2 units/mL test soln.) × (10 mL test soln./1 mL enzyme soln.) × (1 mL enzyme soln./1 mg enzyme powder)
= 2 units/mg enzyme powder.

Objectives

To measure the activity of an enzyme of the citric acid cycle, isocitrate dehydrogenase, and the effect of enzyme concentration on the rate of reaction.

Procedure

1. Turn *on* the spectrophotometer and let it warm up for a few minutes. Turn the wavelength control knob to read 340 nm. With *no sample tube in the sample compartment*, adjust the amplifier control knob so that 0% transmittance or infinite absorbance is read.

2. Prepare a cocktail of reactants in the following manner: In a 10-mL test tube, mix 2.0 mL phosphate buffer, 1.0 mL $MgCl_2$ solution, 1.0 mL 15 mM isocitrate solution, and 5 mL distilled water.

3. To prepare a **Blank** for the spectrophotometric reading, take a sample tube and add to it 1.0 mL of reagent cocktail (prepared as above), 0.2 mL $NADP^+$ solution, and 1 mL of distilled water. Mix the solutions by shaking the sample tube. *Be careful to pipet exactly 0.2 mL $NADP^+$.*

4. Insert the sample tube with the **Blank** solution into the spectrophotometer. Adjust the reading to 100% transmittance (or 0 absorbance). This zeroing must be performed every

10 min. before each enzyme activity run because some instruments have a tendency to drift. The instrument is now ready to measure enzyme activity.

5. Prepare one sample tube for the **enzyme activity measurements**. Add 1.0 mL of reagent cocktail and 0.7 mL of distilled water. Next, add 0.2 mL NADP$^+$ solutions. *Be careful to pipet exactly 0.2 mL NADP$^+$*. Mix the contents of the sample tube. Readjust the spectrophotometer with the "Blank" (prepared in step no. 3) to read 0.00 absorbance or 100.00% transmission. Remove the "Blank" and save it for future readjustments. *In the next step the timing is very important.* Take a watch, and *at a set time* (for example 2 hr. 15 min. 00 sec.) add *exactly* 0.3 mL enzyme solution to the sample tube. Mix it thoroughly and quickly by shaking the tube. Insert the sample tube into the spectrophotometer and take a first reading 1 min. after the mixing time (i.e., 2 hr. 16 min. 00 sec.). Record the absorbancies on your Report Sheet in column 1. Thereafter, take a reading of the spectrophotometer every 30 sec. and record the readings for 5–6 min. on your Report Sheet in column 1.

6. Repeat the experiment exactly as in step no. 5: Preparing the sample solution, readjusting the instrument with the "Blank," and reading the sample solution every 30 sec. for 5 min. Record the spectrophotometric readings on your Report Sheet in column 2.

7. Prepare a new sample tube with the following contents: 1.0 mL reagent cocktail, 0.8 mL distilled water, and 0.2 mL NADP$^+$ solution. *Be careful to pipet exactly 0.2 mL NADP$^+$.* Mix it thoroughly. Readjust the spectrophotometer with the "Blank" to zero absorbance (100% transmission). *At a set time* (i.e., 2 hr. 33 min. 00 sec.), add *exactly* 0.2 mL of enzyme solution. Mix the sample tube and insert into the spectrophotometer. Take your first reading 1 min. after the mixing and every 30 sec. for 5 min. thereafter. Record the absorbancies on your Report Sheet in column 3.

8. Prepare a new sample tube with the following contents: 1.0 mL reagent cocktail, 0.6 mL distilled water, and *exactly* 0.2 mL NADP$^+$ solution. Mix it thoroughly. Readjust the spectrophotometer with the "Blank" to zero absorbance (100% transmission). *At a set time* (i.e., 2 hr. 45 min. 00 sec.), add *exactly* 0.4 mL enzyme solution to the sample tube. Mix it thoroughly. Insert the sample tube into the spectrophotometer. Take a first reading 1 min. after mixing and every 30 sec. for 5 min. thereafter. Record the absorbancies on your Report Sheet in column 4.

9. Plot the numerical data you recorded in the four columns on graph paper. Note that somewhere *between 3 and 5 min.* your graphs are linear. Obtain the slopes of these linear portions and record them on your Report Sheet. Calculate the activities of your enzyme first as (a) units per mL sample solution and second as (b) units per mg enzyme powder.

Chemicals and Equipment

1. Phosphate buffer, pH 7.0
2. 0.1 M $MgCl_2$ solution
3. 6.0 mM $NADP^+$ solution
4. 15.0 mM isocitrate solution
5. Isocitrate dehydrogenase (0.2 mg powder/mL solution)
6. Spectrophotometers with 5 cuvettes each
7. 1-mL graduated pipets

Experiment 24

PRE-LAB QUESTIONS

1. Explain the difference between the structures of citrate and isocitrate.

2. In this experiment you follow the change in the reduced coenzyme concentration by its absorbance at 340 nm. Look up Section 9.2 in your textbook. What part of the electromagnetic spectrum is identified with this wavelength?

3. What other reactant or product concentration could be used to measure the isocitrate dehydrogenase activity?

4. What is the difference between NAD^+ and $NADP^+$? (Consult your textbook.)

Experiment 24

REPORT SHEET

Time (sec.) after mixing	Column 1	Absorbance of sample 2	3	4
60				
90				
120				
150				
180				
210				
240				
270				
300				
330				
360				

1. Plot your data: Absorbance versus time.

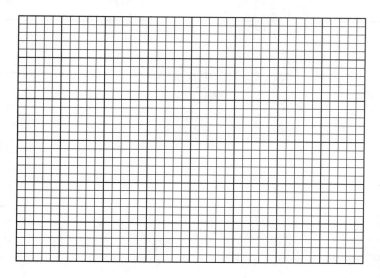

2. Calculate the enzyme activity:

   (a) Units of enzyme activity/mL reaction mixture: The slope of the plot is usually a
   straight line. If so, read the value of change in absorbance per min. Divide it by
   0.01. This gives you the number of enzyme activity units/reaction mixture.
   (One unit of enzyme activity is 0.01 absorbance/min.) Your reaction mixture
   had a volume of 2.2 mL. Thus dividing by 2.2 will give you the activities in
   units/mL reaction mixture.

	(1)	(2)	(3)	(4)
Units/reaction mixture				
Units/mL reaction mixture				

   (b) Calculate the isocitrate dehydrogenase activity per mg powder extract. For
   example, your enzyme solution contained 0.2 mg powder extract/mL solution.
   If you added 0.2 mL of enzyme solution, it contained 0.04 mg powder extract.
   Dividing the units/reaction mixture (obtained above) by the number of mg of
   powder extract added gives you the units/mg powder extract.

	(1)	(2)	(3)	(4)
Units of enzyme activity/mg powder extract				

POST-LAB QUESTIONS

1. You ran two experiments with the same enzyme concentration in (1) and (2). Calculate the average activity for this concentration of enzyme in units/mg powder. Does the reproducibility fall within $\pm 5\%$?

2. Does your enzyme activity (units/mg powder) give you the same number for the different enzyme concentrations employed?

3. If your powder extract contained 80% protein, what would be the average isocitrate dehydrogenase activity per mg protein (enzyme)?

4. In step 2 of your procedure you added $MgCl_2$ to the reaction mixture. What was the purpose of this addition? If you would have forgotten to add this reagent, would the activity of the enzyme be different from that you obtained? If so, in what way?

5. In procedure 3 you prepared a **Blank**. What is a blank?

Quantitative analysis of vitamin C contained in foods

Background

Ascorbic acid is commonly known as vitamin C. It was one of the first vitamins that played a role in establishing the relationship between a disease and its prevention by proper diet. The disease *scurvy* has been known for ages, and a vivid description of it was given by Jacques Cartier, a 16th century explorer of the American continent: "Some did lose their strength and could not stand on their feet. . . . Others . . . had their skin spotted with spots of blood . . . their mouth became stinking, their gums so rotten that all the flesh did fall off." Prevention of scurvy can be obtained by eating fresh vegetables and fruits. The active ingredient in fruits and vegetables that helps to prevent scurvy is ascorbic acid. It is a powerful biological antioxidant (reducing agent). It helps to keep the iron in the enzyme, prolyl hydroxylase, in the reduced form and, thereby, it helps to maintain the enzyme activity. Prolyl hydroxylase is essential for the synthesis of normal collagen. In scurvy, the abnormal collagen causes skin lesions and broken blood vessels.

Vitamin C cannot be synthesized in the human body and must be obtained from the diet (e.g., citrus fruits, broccoli, turnip greens, sweet peppers, tomatoes) or by taking synthetic vitamin C (e.g., vitamin C tablets, "high-C" drinks, and other vitamin C-fortified commercial foods). The minimum recommended adult daily requirement of vitamin C to prevent scurvy is 60 mg. Some people, among them the late Linus Pauling, twice Nobel Laureate, suggested that very large daily doses (250 to 10,000 mg) of vitamin C could help prevent the common cold, or at least lessen the symptoms for many individuals. No reliable medical data support this claim. At present, the human quantitative requirement for vitamin C is still controversial and requires further research.

In this experiment, the amount of vitamin C is determined quantitatively by titrating the test solution with a water-soluble form of iodine I_3^-:

$$:\ddot{I}:\ddot{I}: + :\ddot{I}:^- \longrightarrow \left[:\ddot{I}:\ \ddot{I}::\ddot{I}: \right]^- \text{(tri-iodode ion)}$$

Expanded octets

Vitamin C is oxidized by I_2 (as I_3^-) according to the following chemical reaction:

Vitamin C	(MW 254)	Oxidized product
(MW 176)		(MW 174)

As vitamin C is oxidized by iodine, I_2 becomes reduced to I^-. When the end point is reached (no vitamin C is left), the excess of I_2 will react with a starch indicator to form a starch-iodine complex which is blackish-blue in color.

$$I_2 + starch \rightarrow iodine - starch\ complex\ (blackish\text{-}blue)$$

It is worthwhile to know that although vitamin C is very stable when dry, it is readily oxidized by air (oxygen) when in solution; therefore, a solution of vitamin C should not be exposed to air for long periods. The amount of vitamin C can be calculated by using the following conversion factor:

$$1\ mL\ of\ I_2\ (0.01\ M) = 1.76\ mg\ vitamin\ C$$

Objective

To determine the amount of vitamin C that is present in certain commercial food products by the titration method.

Procedure

1. Pour about 60 mL of a fruit drink that you wish to analyze into a clean, dry 100-mL beaker. The fruit drink should be light colored, apple, orange, or grapefruit, but not dark colored, such as grape. Record the kind of drink on the Report Sheet (1).

2. If the fruit drink is cloudy or contains suspended particles, it can be clarified by the following procedure: Add Celite, used as a filter aid, to the fruit drink (about 0.5 g). After swirling it thoroughly, filter the solution through a glass funnel, bedded with a large piece of cotton. Collect the filtrate in a 50-mL Erlenmeyer flask (Fig. 25.1).

3. Using a 10-mL volumetric pipet and a Spectroline pipet filler, transfer 10.00 mL of the fruit drink into a 125-mL Erlenmeyer flask. Then add 20 mL of distilled water, 5 drops of 3 M HCl (as a catalyst), and 10 drops of 2% starch solution to the flask.

4. Clamp a clean, dry 50-mL buret onto the buret stand. Rinse the buret twice with 5-mL portions of iodine solution. Let the rinses run through the tip of the buret and discard them. Fill the buret slightly above the zero mark with a standardized iodine solution. (A dry funnel may be used for easy transfer.) Air bubbles should be removed by turning

the stopcock several times to force the air bubbles out of the tip. Record the molarity of standardized iodine solution (2). Record the initial reading of standardized iodine solution to the nearest 0.02 mL (3a).

Figure 25.1
Clarification of fruit drinks.

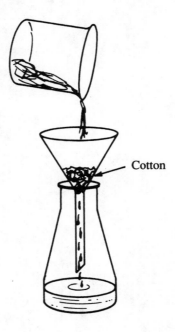

Cotton

5. Place the flask that contains the vitamin C sample under the buret and add the iodine solution dropwise, while swirling, until the indicator just changes to dark blue. This color should persist for at least 20 sec. Record the final buret reading (3b). Calculate the total volume of iodine solution required for the titration (3c), the weight of vitamin C in the sample (4), and percent (w/v) of vitamin C in the drink (5). Repeat this titration procedure twice more, except using 20- and 30-mL portions of the same fruit drink instead of 10 mL. Record the volumes of iodine solution that are required for each titration.

Chemicals and Equipment

1. 50-mL buret
2. Buret clamp
3. Spectroline pipet filler
4. 10-mL volumetric pipet
5. 50-mL Erlenmeyer flask
6. Cotton
7. Filter aid
8. Hi-C apple drink
9. Hi-C orange drink
10. Hi-C grapefruit drink
11. 0.01 M iodine in potassium iodide
12. 3 M HCl
13. 2% starch solution

Experiment 25

PRE-LAB QUESTIONS

1. What are the symptoms of scurvy?

2. Vitamin C is also called ascorbic acid. Write the structure of vitamin C. Where do you find an acid group? Circle it in the structure.

3. What is the minimum daily requirement of vitamin C to prevent scurvy in adults?

4. What enzyme is oxidized in the absence of vitamin C and causes the symptoms of scurvy? Which natural product is synthesized by this enzyme?

Experiment 25

REPORT SHEET

1. The kind of fruit drink _____

2. Molarity of iodine solution _____

3. Titration results

	Sample 1 (10.0 mL)	Sample 2 (20.0 mL)	Sample 3 (30.0 mL)
a. Initial buret reading	_____ mL	_____ mL	_____ mL
b. Final buret reading	_____ mL	_____ mL	_____ mL
c. Total volume of iodine solution used: (b − a)	_____ mL	_____ mL	_____ mL

4. The weight of vitamin C in the fruit drink sample: [(3c) × 1.76 mg/mL]   _____ mg   _____ mg   _____ mg

5. Concentration of vitamin C in the fruit drink (mg/100 mL): [(4)/volume of drink] × 100   _____   _____   _____

6. Average concentration of vitamin C in the fruit drink _____ mg/100 mL

POST-LAB QUESTIONS

1. Why is HCl added for the titration of vitamin C?

2. What gives the blue color in your titration?

3. What volume of fruit drink would satisfy your minimum daily vitamin C requirement?

4. Why was it necessary to use Celite (filter aid)?

5. You have analyzed a 15.0-mL sample of orange juice for vitamin C. Using a 0.005 M iodine solution for titration, your initial buret reading was 1.0 mL and the final reading was 7.4 mL. What was the concentration of vitamin C (mg/100 mL) in your orange juice? Show your work.

Experiment 26

Analysis of vitamin A in margarine

Background

Vitamin A, or retinol, is one of the major fat-soluble vitamins. It is present in many foods; the best natural sources are liver, butter, margarine, egg yolk, carrots, spinach, and sweet potatoes. Vitamin A is the precursor of retinal, the essential component of the visual pigment rhodopsin.

Vitamin A (All-*trans*-retinol)   11-*cis*-retinal

When a photon of light penetrates the eye, it is absorbed by the 11-*cis*-retinal. The absorption of light converts the 11-*cis*-retinal to all-*trans*-retinal:

11-*cis*-retinal   All-*trans*-retinal

This isomerization converts the energy of a photon into an atomic motion which in turn is converted into an electrical signal. The electrical signal generated in the retina of the eye is transmitted through the optic nerve into the brain's visual cortex.

Even though part of the all-*trans*-retinal is regenerated in the dark to 11-*cis*-retinal, for good vision, especially for night vision, a constant supply of vitamin A is needed. The recommended daily allowance of vitamin A is 750 μg. Deficiency in vitamin A results in night blindness and keratinization of epithelium. The latter compromises the integrity of healthy skin. In young animals, vitamin A is also required for growth. On the other hand, large doses of vitamin A, sometimes recommended in faddish diets, can be harmful. A daily dose above 1500 μg can be toxic

To analyze the vitamin A content of margarine by spectrophotometric method.

Procedure

The analysis of vitamin A requires a multistep process. In order that you should be able to follow the step-by-step procedure, a flow chart is provided here:

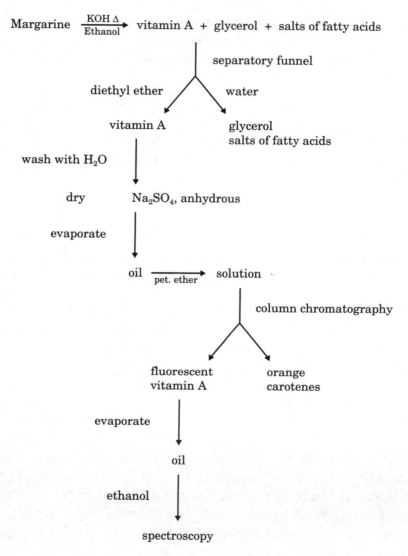

1. Margarine is largely fat. In order to separate vitamin A from the fat in margarine, first the sample must be saponified. This converts the fat to water-soluble products, glycerol and potassium salts of fatty acids. Vitamin A can be extracted by diethyl ether from the products of the saponification process. To start, weigh a cover glass to the nearest 0.1 g. Report this weight on the Report Sheet (1). Add approximately 10 g of margarine to the watch glass. Record the weight of watch glass plus sample to the nearest 0.1 g on your

Report Sheet (2). Transfer the sample from the watch glass into a 250-mL Erlenmeyer flask with the aid of a glass rod, and wash it in with 75 mL of 95% ethanol. Add 25 mL of 50% KOH solution. Cover the Erlenmeyer flask loosely with a cork and put it on an electric hot plate. Bring it gradually to a boil. Maintain the boiling for 5 min. with an occasional swirling of the flask using tongs. The stirring should aid the complete dispersal of the sample. Remove the Erlenmeyer from the hot plate and let it cool to room temperature (approximately 20 min.).

CAUTION!

50% KOH solution can cause burns on your skin. Handle the solution with care, do not spill it. If a drop gets on your skin, wash it immediately with copious amounts of water. Use gloves when working with this solution.

2. While the sample is cooling, prepare a chromatographic column. Take a 25-mL buret. Add a small piece of glass wool. With the aid of a glass rod, push it down near the stopcock. Add 15–16 mL of petroleum ether to the buret. Open the stopcock slowly, and allow the solvent to fill the tip of the buret. Close the stopcock. You should have 12–13 mL of petroleum ether above the glass wool. Weigh about 20 g of alkaline aluminum oxide (alumina) in a 100-mL beaker. Place a small funnel on top of your buret. Pour the alumina slowly, in small increments, into the buret. Allow it to settle to form a 20-cm column. Drain the solvent but do not allow the column to run dry. **Always have at least 0.5 mL clear solvent on top of the column.** If the alumina adheres to the walls of the buret, wash it down with more solvent.

CAUTION!

Diethyl ether is very volatile and flammable. Make certain that there are no open flames, not even a hot electrical plate in the vicinity of the operation.

3. Transfer the solution (from your reaction in step no. 1) from the Erlenmeyer flask to a 500-mL separatory funnel. Rinse the flask with 30 mL of distilled water and add the rinsing to the separatory funnel. Repeat the rinsing two more times. Add 100 ml of diethyl ether to the separatory funnel. Close the separatory funnel with the glass stopper. Shake the separatory funnel vigorously. (See Exp. 12 Fig. 12.1 for technique.) Allow it to separate into two layers. Drain the bottom aqueous layer into an Erlenmeyer flask. Add the top (diethyl ether) layer to a second clean 250-mL Erlenmeyer flask. Pour back the aqueous layer into the separatory funnel. Add another 100-mL portion of diethyl ether. Shake and allow it to separate into two layers. Drain again the bottom (aqueous) layer and discard. Combine the first diethyl ether extract with the residual diethyl ether extract in the separatory funnel. Add 100 mL of distilled water to the combined diethyl ether extracts in the separatory funnel. Agitate it gently and allow the water to drain. Discard the washing.

4. Transfer the diethyl ether extracts into a clean 300-mL beaker. Add 3–5 g of anhydrous Na_2SO_4 and stir it gently for 5 min. to remove traces of water. Decant the diethyl ether extract into a clean 300-mL beaker. Add a boiling chip or a boiling stick. Evaporate the diethyl ether solvent to about 25 mL volume by placing the beaker **in the hood** on a steam bath. Transfer the sample to a 50-mL beaker and continue to evaporate on the steam bath until an oily residue forms. Remove the beaker from the steam bath. Cool it in an ice bath for 1 min. Add 5 mL of petroleum ether and transfer the liquid (without the boiling chip) to a 10-mL volumetric flask. Add sufficient petroleum ether to bring it to volume.

5. Add 5 mL of extracts in petroleum ether to your chromatographic column. By opening the stopcock drain the sample into your column, but **take care not to let the column run dry**. (Always have about 0.5 mL liquid on top of the column.) Continue to add solvent to the top of your column. Collect the eluents in a beaker. First you will see the orange-colored carotenes moving down the column. With the aid of a UV lamp, you can also observe a fluorescent band following the carotenes. This fluorescent band contains your vitamin A. Allow all the orange color band to move to the bottom of your column and into the collecting beaker. When the fluorescent band reaches the bottom of the column, close the stopcock. By adding petroleum ether on the top of the column continuously, elute the fluorescent band from the column into a 25-mL graduated cylinder. Continue the elution until all the fluorescent band has been drained into the graduated cylinder. Close the the stopcock, and record the volume of the eluate in the graduated cylinder on your Report Sheet (4). Add the vitamin A in the petroleum ether eluate to a dry and clean 50-mL beaker. Evaporate the solvent in the hood on a steam bath. The evaporation is complete when an oily residue appears in the beaker. Add 5 mL of absolute ethanol to the beaker. Transfer the sample into a 10-mL volumetric flask and bring it to volume by adding absolute ethanol.

6. Place your sample in a 1-cm length quartz spectroscopic cell. The control (blank) spectroscopic cell should contain absolute ethanol. Read the absorbance of your sample against the blank, according to the instructions of your spectrophotometer, at 325 nm. Record the absorption at 325 nm on your Report Sheet (5).

7. Calculate the amount of margarine that yielded the vitamin A in the petroleum ether eluate. Remember that you added only half (5 mL) of the extract to the column. Report this value on your Report Sheet (6). Calculate the grams of margarine that would have yielded the vitamin A in 1 mL absolute ethanol by dividing (6)/10 mL. Record it on your Report Sheet (7). Calculate the vitamin A in a pound of margarine by using the following formula:

$$\mu g \text{ vitamin A/lb of margarine} = \text{Absorption} \times 5.5 \times [454/(7)].$$

Record your value on the Report Sheet (8).

Chemicals and Equipment

1. Separatory funnel (500 mL)
2. Buret (25 mL)
3. UV lamp
4. Spectrophotometer (near UV)
5. Margarine
6. Petroleum ether (30−60°C)
7. 95% ethanol
8. Absolute ethanol
9. Diethyl ether
10. Glass wool
11. Alkaline aluminum oxide (alumina)

Experiment 26

PRE-LAB QUESTIONS

1. The structure of β-carotene is given below. What is the difference between β-carotene and vitamin A?

β-carotene

2. Why must you work in the hood when dealing with diethyl ether solutions? Why should you make sure that there is no lit Bunsen burner in the lab during this experiment?

3. In the saponification process, you hydrolyzed fat in the presence of KOH. Write an equation of a reaction in which the fat is hydrolyzed in the presence of HCl. What is the difference between the products of the saponification and that of the acid hydrolysis?

4. There is a warning in the procedures regarding the use of 50% KOH. Why is this solution so dangerous? (See Box 8B in your textbook.)

Experiment 26

REPORT SHEET

1. Weight of watch glass   _____ g

2. Weight of watch glass + margarine   _____ g

3. Weight of margarine: (2) − (1)   _____ g

4. Volume of petroleum ether eluate   _____ mL

5. Absorption at 325 nm   _____

6. Grams margarine in 1 mL of petroleum ether eluate: 2 × [(3)/(4)]   _____ g

7. Grams of margarine in 1 mL of absolute ethanol: (6)/10 mL   _____ g

8. μg vitamin A/lb margarine: (5) × 5.5 × [454/(7)]   _____

POST-LAB QUESTIONS

1. How did you detect the fluorescent band of vitamin A during the chromatography? Was it easy to see?

2. In your separation scheme the fatty acids of the margarine ended up in the aqueous wash, which was discarded. Could you have removed the fatty acids, similarly, if, instead of saponification, you used acid hydrolysis? Explain.

3. What chemical processes are needed to convert vitamin A to the 11-*cis*-retinal?

4. The label on a commercial margarine sample states that 1 g of it contains 15% of the daily recommended allowance. Was your sample richer or poorer in vitamin A than the above mentioned commercial sample?

Experiment 27

Urine analysis

Background

The kidney is an important organ that filters materials from the blood that are harmful, or in excess, or both. These materials are excreted in the urine. A number of tests are routinely run in clinical laboratories on urine samples. These involve the measurements of glucose or reducing sugars, ketone bodies, albumin, specific gravity, and pH.

Normal urine contains little or no glucose or reducing sugars; the amount varies from 0.05 to 0.15%. Higher concentrations may occur if the diet contains a large amount of carbohydrates or if strenuous work was performed shortly before the test. Patients with diabetes or liver damage have chronically elevated glucose content in the urine. A semiquantitative test of glucose levels can be performed with the aid of test papers such as Clinistix. This is a quick test that uses a paper containing two enzymes—glucose oxidase and peroxidase. In the presence of glucose, the glucose oxidase catalyzes the formation of gluconic acid and hydrogen peroxide. The hydrogen peroxide is decomposed with the aid of peroxidase and yields atomic oxygen.

α-D-glucose   Gluconic acid   Hydrogen peroxide

$$H_2O_2 \xrightarrow{\text{Peroxidase}} H_2O + [O]$$

In Clinistix, the atomic oxygen reacts with an indicator, o-tolidine, and produces a purple color. The intensity of the purple color is proportional to the glucose concentration.

o-Tolidine

This test is specific for glucose only. No other reducing sugar will give positive results.

Normal urine contains no albumin or only a trace amount of it. In case of kidney failure or malfunction, the protein passes through the glomeruli and is not reabsorbed in the tubule. So, albumin and other proteins end up in the urine. The condition known as proteinuria may be symptomatic of kidney disease. The loss of albumin and other blood proteins will decrease the osmotic pressure of blood. This allows water to flow from the blood into the tissues, and creates swelling (edema). Renal malfunction is usually accompanied by swelling of the tissues. The Albustix test is based on the fact that a certain indicator at a certain pH changes its color in the presence of proteins.

Albustix contains the indicator, tetrabromophenol blue, in a citrate buffer at pH 3. At this pH, the indicator has a yellow color. In the presence of protein, the color changes to green. The higher the protein concentration, the greener the indicator will be. Therefore, the color produced by the Albustix can be used to estimate the concentration of protein in urine.

Three substances that are the products of fatty acid catabolism—acetoacetic acid, β-hydroxybutyric acid, and acetone—are commonly called ketone bodies. These are normally present in the blood in small amounts and can be used as an energy source by the cells. Therefore, no ketone bodies will normally be found in the urine. However, when fats are the only energy source, excess production of ketone bodies will occur. They will be filtered out by the kidney and appear in the urine. Such abnormal conditions of high fat catabolism take place during starvation or in diabetes mellitus when glucose, although abundant, cannot pass through the cell membranes to be utilized inside where it is needed. Acetoacetic acid (CH_3COCH_2COOH), and to a lesser extent acetone (CH_3COCH_3) and β-hydroxybutyric acid ($CH_3CHOHCH_2COOH$), react with sodium nitroprusside $\{Na_2[Fe(CN)_5NO]\}\cdot2H_2O$ to give a maroon-colored complex. In Ketostix, the test area contains sodium nitroprusside and a sodium phosphate buffer to provide the proper pH for the reaction. The addition of lactose to the mixture in the Ketostix enhances the development of the color.

Some infants are born with a genetic defect known as phenylketonuria (PKU). They lack the enzyme phenylalanine oxidase, which converts phenylalanine to tyrosine. Thus phenylalanine accumulates in the body and it is degraded to phenylpyruvate by transamination:

| Phenylalanine | Pyruvate | Phenylpyruvate | Alanine |

Phenylpyruvate is excreted in the urine. Normal urine does not contain any phenylpyruvate. People suffering from PKU have varying amounts of phenylpyruvate in their urine. PKU causes severe mental retardation in infants if it is not treated immediately after birth, which is done by restricting the phenylalanine content of the diet. In many states, the law requires that every newborn be tested for phenylpyruvate in the urine. The test is based on the reaction of the iron (III) ion with the phenylpyruvate, producing a gray-green color.

Phenistix strips which are coated with $Fe(NH_4)(SO_4)_2$ and a buffer can detect as little as 8 mg of phenylpyruvate in 100 mL of urine. Some drugs such as aspirin produce metabolites (salicylic acid) that are excreted in the urine and give color with an iron(III) ion. However, this produces a deep purple color and not the gray-green of PKU. The purple color that is given by the Phenistix can be used to diagnose an overdose of aspirin. Other drugs, such as phenylthiazine, in an overdose, give a gray-purple color with Phenistix. For PKU diagnosis only the appearance of the gray-green color means a positive test.

Urobilinogen and other bile pigments are normally minor components of urine (2 to 50 μg/100 mL). They are the products of hemoglobin breakdown. Bile pigments are usually excreted in the feces. In case of obstruction of the bile ducts (gallstones, obstructive jaundice), the normal excretion route through the small intestines is blocked and the excess bilirubin is filtered out of the blood by the kidneys and appears in the urine. Urobilistix is a test paper that can detect the presence of urobilinogen, because it is impregnated with p-dimethylaminobenzaldehyde. In strongly acidic media, this reagent gives a yellow-brownish color with urobilinogen.

p-Dimethylaminobenzaldehyde

The specific gravity of normal urine may range from 1.008 to 1.030. After excessive fluid intake (like a beer party), the specific gravity may be on the low side; after heavy exercise and perspiration, it may be on the high side. High specific gravity indicates excessive dissolved solutes in the urine.

The pH of normal urine can vary between 4.7 through 8.0. The usual value is about 6.0. High-protein diets and fever can lower the pH of urine. In severe acidosis, the pH may be as low as 4.0. Vomiting and respiratory or metabolic alkalosis can raise the pH above 8.0.

Objectives

1. To perform quick routine analytical tests on urine samples.
2. To compare results obtained on "normal" and "pathological" urine samples.

Procedure

Each student must analyze her (his) own urine. A fresh urine sample will be collected prior to the laboratory period. The stockroom will provide paper cups for sample collection. *While handling body fluids, such as urine, plastic gloves should be worn.* The used body fluid will be collected in a special jar and disposed of collectively. The plastic gloves worn will be collected and autoclaved before disposal. In addition, the stockroom will provide one "normal" and two "pathological" urine samples.

Place 5 mL of the urine sample from each source into four different test tubes. These will be tested with the different test papers.

Glucose Test

For the glucose test, use for comparison two test tubes, each half-filled, one with 0.25% and the other with 1% glucose solutions. Take six strips of Clinistix from the bottle. Replace the cap immediately. Dip the test area of the Clinistix into one of the samples. Tap the edge of the strip against a clean, dry surface to remove the excess urine. Compare the test area of the strip to the color chart supplied on the bottle exactly 10 sec. after the wetting. Do not read the color changes that occur after 10 sec. Record your observation: The "light" on the color chart means 0.25%, or less, glucose; the "medium" means 0.4%; and the "dark" means 0.5%, or more. Repeat the test with the other five samples.

Protein Test

For the protein test, take four Albustix strips, one for each of the four urine samples, from the bottle. Replace the cap immediately. Dip the Albustix strips into the test solutions, making certain that the reagent area of the strip is completely immersed. Tap the edge of the strip against a clean, dry surface to remove the excess urine. Compare the color of the Albustix test area with the color chart supplied on the bottle. The time of the comparison is not critical; you can do it immediately or any time within 1 min. after wetting. Read the color from yellow (negative) to different shades of green, indicating trace amounts of albumin, up to 0.1%.

Ketone Bodies

To measure the ketone bodies' concentration in the urine, take four Ketostix strips from the bottle. Replace the cap immediately. Dip a Ketostix in each of the urine samples and remove the strips at once. Tap them against a clean, dry surface to remove the excess urine. Compare the color of the test area of the strips to the color chart supplied on the bottle. Read the colors exactly 15 sec. after wetting. A buff pink color indicates the absence of ketone bodies. A progression to a maroon color indicates increasing concentration of ketone bodies from 52 to 160 mg/L of urine.

Test for PKU

To test for PKU disease, use four Phenistix strips, one for each urine sample. Immerse the test area of the strips into the urine samples and remove them immediately. Remove the excess urine by tapping the strips against a clean, dry surface. Read the color after 30 sec. of wetting, and match them against the color charts provided on the bottle. Record your estimated phenylpyruvate content: negative or 0.015 to 0.1%.

Urobilinogen

To measure the urobilinogen content of the urine samples, use Urobilistix strips, one for each urine sample. Dip them into the samples and remove the excess urine by tapping the strips against a clean, dry surface. Sixty seconds after wetting, compare the color of the test area of the strips to the color chart provided on the bottle. Estimate the urobilinogen content and record it in Ehrlich units.

pH in Urine

To measure the pH of each urine sample, use pH indicator paper such as pHydrion test paper within the pH range of 3.0 to 9.0.

For the preceding tests, you may use a multipurpose strip such as Labstix that contains test areas for all these tests, except for the test for phenylpyruvate, on one strip. The individual test areas are separated from each other and clearly marked. The time requirements to read the colors are also indicated on the chart. The results of five tests can be read in 1 min.

Specific Gravity of Urine

The specific gravity of your urine samples will be measured with the aid of a hydrometer (urinometer; see Fig. 27.1). Place the bulb in a cylinder. Add sufficient urine to the cylinder to make the bulb float. Read the specific gravity of the sample from the stem of the hydrometer where the meniscus of the urine intersects the calibration lines. Be sure the hydrometer is freely floating and does not touch the walls of the cylinder. In order to use as little urine as possible, the instructor may read the normal and two pathological urine samples for the whole class. If so, you will measure the specific gravity of your own urine sample only.

Figure 27.1
A urinometer.

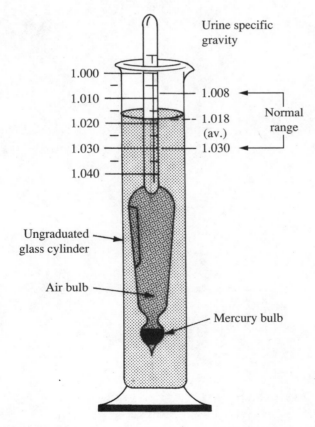

Chemicals and Equipment

1. 0.25 and 1% glucose solutions
2. Clinistix
3. Albustix
4. Ketostix
5. Urobilistix
6. pH paper in the 3.0 to 9.0 range
7. A multipurpose Labstix (instead of these test papers)
8. Phenistix
9. 3 M NaOH
10. Hydrometer (urinometer)

NAME SECTION DATE

PARTNER GRADE

Experiment 27

PRE-LAB QUESTIONS

1. Why should you wear gloves when dealing with urine samples?

2. In what tests do we use the following reagents?

a. Fe^{3+}

b.

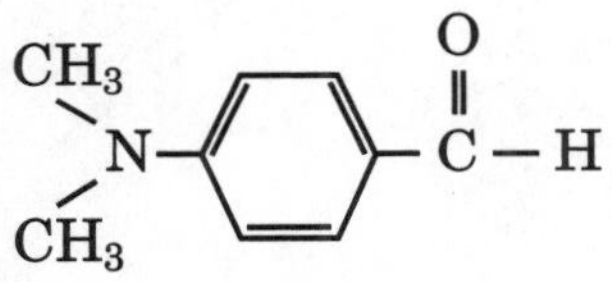

c.

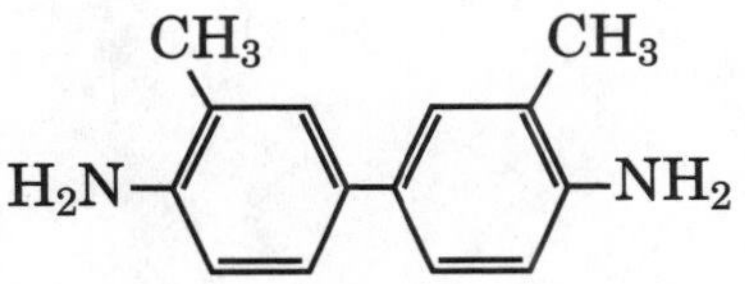

3. What does a high specific gravity reading on a urine sample indicate?

4. A patient's urine shows a high specific gravity, 1.04. The pH is 7.8, and the Phenistix test indicates a purple color that is not characteristic of PKU. The patient has had a high fever for a few days and has been given aspirin. Do these tests indicate any specific disease, or are they symptomatic of recovering from a high fever? Explain.

Experiment 27

REPORT SHEET

Urine Samples

Test	Normal	Pathological A	Pathological B	Your own	Remarks
Glucose					
Ketone bodies					
Albumin					
Urobilinogen					
pH					
Phenylpyruvate					
Specific gravity					

POST-LAB QUESTIONS

1. Did you find any indication that your urine is not normal? If so, what may be the reason?

2. Why is the phenylpyruvate test mandatory with newborns in many states?

3. If a urine sample shows unusually high protein content, what disease is suggested by this test?

4. A patient's urine was tested with Clinistix, and the color was read 60 sec. after wetting the strip. It showed 1.0% glucose in the urine. Is the patient diabetic? Explain.

5. Assume that you did not have enough urine to completely immerse the urinometer. Would your specific gravity readings be meaningful? Explain your answer.

Harcourt, Inc.

List of Apparatus and Equipment in Student's Locker

Amount and description

(1) Beaker, 50 mL
(1) Beaker, 100 mL
(1) Beaker, 250 mL
(1) Beaker, 400 mL
(1) Beaker, 600 mL
(1) Clamp, test tube
(1) Cylinder, graduated by 0.1 mL, 10 mL
(1) Cylinder, graduated by 1 mL, 100 mL
(1) Dropper, medicine with rubber bulb
(1) Evaporating dish
(1) Flask, Erlenmeyer, 125 mL
(1) Flask, Erlenmeyer, 250 mL
(1) Flask, Erlenmeyer, 500 mL
(1) File, triangular
(1) Forceps
(1) Funnel, short stem
(1) Gauze, wire
(1) Spatula, stainless steel
(1) Sponge
(1) Striker (or box of matches)
(6) Test tubes, approximately 15 × 150 mm
(1) Test tube brush
(1) Thermometer, 150°C
(1) Tongs, crucible
(1) Wash bottle, plastic
(1) Watch glass

Appendix 2

List of Common Equipment and Materials in the Laboratory

Each laboratory should be equipped with hoods and safety-related items such as fire extinguisher, fire blankets, safety shower, and eye wash fountain. The equipment and materials listed here for 25 students should be made available in each laboratory.

Acid tray
Aspirators (splashgun type) on sink faucet
Balances, single pan, triple beam (or Centogram) or top-loading
Barometer
Clamps, extension
Clamps, thermometer
Clamps, utility
Containers for solid chemical waste disposal
Containers for liquid organic waste disposal
Corks
Detergent for washing glassware
Drying oven
Filter paper
Glass rods, 4 and 6 mm OD
Glass tubing, 6 and 8 mm OD
Glycerol (glycerine) in dropper bottles
Hot plates
Ice maker
Paper towel dispensers
Pasteur pipets
Rings, support, iron, 76 mm OD
Ring stands
Rubber tubing, pressure
Rubber tubing, latex (0.25 in. OD)
Water, deionized or distilled
Weighing dishes, polystyrene, disposable, $73 \times 73 \times 25$ mm
Weighing paper

Special Equipment and Chemicals

In the instructions below every time a solution is to be made up in "water" you must use *distilled water*.

Experiment 1   Structure in organic compounds: use of molecular models. I

Special Equipment

(Color of spheres may vary depending on the set; substitute as necessary.)

(50)	Black spheres—4 holes
(300)	Yellow spheres—1 hole
(50)	Colored spheres (e.g. green)—1 hole
(25)	Blue spheres—2 holes
(400)	Sticks
(25)	Protractors
(75)	Springs (optional)

Experiment 2   Stereochemistry: use of molecular models. II

Special Equipment

Commercial molecular model kits vary in style, size, material composition, and the color of the components. The set which works best in this exercise is the *Molecular Model Set for Organic Chemistry* available from Allyn and Bacon, Inc. (Newton, MA). Wood ball and stick models work as well. For 25 students, 25 of these sets should be provided. If you wish to make up your own kit, you would need the following for 25 students:

(25)   Cyclohexane model kits: each consisting of the following components:
8 carbons—black, 4 hole
18 hydrogens—white, 1 hole
2 substituents—red, 1 hole
24 connectors—bonds

(25)   Chiral model kits: each consisting of the following components:
8 carbons—black, 4 hole
32 substituents—8 red, 1 hole; 8 white, 1 hole; 8 blue, 1 hole; 8 green, 1 hole
28 connectors—bonds

(5)   Small hand mirrors

Experiment 3   Identification of hydrocarbons

Special Equipment

(2 vials)   Litmus paper, blue
(250) 100×13 mm test tubes

Chemicals

(25 g)   Iron filings or powder. Clean the iron filings with 3 M HCl before using. Cover the iron filings with 3 M HCl and stir with a glass rod. Gravity filter to remove the solution and wash with water. Blot dry the iron filings with paper towels and dry in an oven.

The following solutions should be placed in dropper bottles.

(100 mL)   Concentrated H_2SO_4 (18 M H_2SO_4)
(100 mL)   Cyclohexene
(100 mL)   Hexane
(100 mL)   Ligroin (b.p. 90–110°C)
(100 mL)   Toluene
(100 mL)   1% Br_2 in cyclohexane **(wear a face shield, rubber gloves, and a rubber apron; prepare under hood)**: mix 1.0 mL Br_2 with enough cyclohexane to make 100 mL. *Prepare fresh solutions prior to use; keep in a dark-brown dropper bottle; do not store.*
(100 mL)   1% aqueous $KMnO_4$: dissolve 1.0 g potassium permanganate in 50 mL distilled water by gently heating for 1 hr.; cool and filter; dilute to 100 mL. Store in a dark-brown dropper bottle.
(100 mL)   Unknown A = hexane
(100 mL)   Unknown B = cyclohexene
(100 mL)   Unknown C = toluene

Experiment 4   Column and paper chromatography; separation of plant pigments

Special Equipment

(50)   Melting point capillaries open at both ends
(25)   25-mL burets
(1 jar)   Glass wool
(25)   Filter papers (Whatman no.1), 20×10 cm
(3)   Heat lamp (optional)
(25)   Ruler with both English and metric scale
(1)   Stapler
(15)   Hot plates with or without water bath

Chemicals

(1 lb)   Tomato paste
(500 g)   Aluminum oxide (alumina)
(500 mL)   95% ethanol
(500 mL)   Petroleum ether, b.p. 30–60°C

(500 mL)	Eluting solvent: mix 450 mL petroleum ether with 10 mL toluene and 40 mL acetone.
(10 mL)	0.5% β-carotene solution: dissolve 50 mg in 10 mL petroleum ether. Wrap the vial in aluminum foil to protect from light and keep in refrigerator until used.
(150 mL)	Saturated bromine water: mix 5.5 g bromine with 150 mL water. **Prepare in hood; wear a face shield, rubber gloves, and a rubber apron.**
(500 mg)	Iodine crystals

Experiment 5   Identification of alcohols and phenols

Special Equipment

(125)	Corks (for test tubes 100×13 mm)
(125)	Corks (for test tubes 150×18 mm)
(25)	Hot plate
(5 rolls)	Indicator paper (pH $1-12$)

Chemicals

The following solutions should be placed in dropper bottles.

(100 mL)	Acetone (reagent grade)
(100 mL)	1-Butanol
(100 mL)	2-Butanol
(100 mL)	2-Methyl-2-propanol (*t*-butyl alcohol)
(100 mL)	Dioxane
(200 mL)	20% aqueous phenol: dissolve 80 g of phenol in 20 mL distilled water; dilute to 400 mL.
(100 mL)	Lucas reagent **(prepare under hood; wear a face shield, rubber gloves, and a rubber apron)**: cool 100 mL of concentrated HCl (12 M HCl) in an ice bath; with stirring, add 150 g anhydrous $ZnCl_2$ to the cold acid.
(150 mL)	Chromic acid solution **(prepare under hood; wear a face shield, rubber gloves, and a rubber apron)**: dissolve 20 g potassium dichromate, $K_2Cr_2O_7$, in 100 mL concentrated sulfuric acid (18 M H_2SO_4). Carefully add this solution to enough ice cold water to bring to 1 L.
(100 mL)	2.5% iron(III) chloride solution: dissolve 2.5 g anhydrous $FeCl_3$ in 50 mL water; dilute to 100 mL.
(100 mL)	Iodine in KI solution: mix 20 g of KI and 10 g of I_2 in 100 mL water
(250 mL)	6 M sodium hydroxide, 6 M NaOH: dissolve 60.00 g NaOH in 100 mL water. Dilute to 250 mL with water.
(100 mL)	Unknown A = 1-butanol
(100 mL)	Unknown B = 2-butanol
(100 mL)	Unknown C = 2-methyl-2-propanol (*t*-butyl alcohol)
(100 mL)	Unknown D = 20% aqueous phenol

Special Equipment

(250)	Corks (to fit 100×13 mm test tube)
(125)	Corks (to fit 150×18 mm test tube)
(1 box)	Filter paper (students will need to cut to size)
(25)	Hirsch funnels
(25)	Hot plates
(25)	Neoprene adapters (no. 2)
(25)	Rubber stopper assemblies: a no. 6 one-hole stopper fitted with glass tubing (15 cm in length $\times$ 7 mm OD)
(25)	50-mL side-arm filter flasks
(25)	250-mL side-arm filter flasks
(50)	Vacuum tubing, heavy-walled (2-ft. lengths)

Chemicals

(50 g)	Hydroxylamine hydrochloride
(100 g)	Sodium acetate

The following solutions should be placed in dropper bottles.

(100 mL)	Acetone (reagent grade)
(100 mL)	Benzaldehyde (freshly distilled)
(100 mL)	*Bis*(2-ethoxymethyl) ether
(100 mL)	Cyclohexanone
(100 mL)	Dioxane
(500 mL)	Ethanol (absolute)
(500 mL)	Ethanol (95%)
(100 mL)	Isovaleraldehyde
(500 mL)	Methanol
(100 mL)	Pyridine
(150 mL)	Chromic acid reagent: dissolve 20 g potassium dichromate, $K_2Cr_2O_7$, in 100 mL concentrated sulfuric acid (18 M H_2SO_4). Carefully add this solution to enough ice cold water to bring to 1 L. **Wear a face shield, rubber gloves, and a rubber apron during the preparation. Do in the hood.**

Tollens' reagent

(100 mL)	Solution A: dissolve 9.0 g silver nitrate in 90 mL of water; dilute to 100 mL.
(100 mL)	Solution B: 10 g NaOH dissolved in enough water to make 100 mL
(100 mL)	10% ammonia water: 35.7 mL of concentrated (28%) NH_3 diluted to 100 mL
(100 mL)	6 M sodium hydroxide, 6 M NaOH: dissolve 24.00 g NaOH in enough water to make 100 mL
(500 mL)	Iodine-KI solution: mix 100 g of KI and a 50 g of iodine in enough distilled water to make 500 mL
(100 mL)	2,4-dinitrophenylhydrazine reagent: dissolve 3.0 g of 2,4-dinitrophenylhydrazine in 15 mL concentrated H_2SO_4 (18 M H_2SO_4). In a beaker, mix together 10 mL water and 75 mL 95% ethanol. With vigorous stirring slowly add

the 2,4-dinitrophenylhydrazine solution to the aqueous ethanol mixture. After thorough mixing, filter by gravity through a fluted filter paper. **Wear a face shield, rubber gloves, and a rubber apron during the preparation. Do in the hood.**

(100 mL)	Semicarbazide reagent: dissolve 22.2 g of semicarbazide hydrochloride in 100 mL of distilled water
(100 mL)	Unknown A = isovaleraldehyde
(100 mL)	Unknown B = benzaldehyde
(100 mL)	Unknown C = cyclohexanone
(100 mL)	Unknown D = acetone

Additional compounds for use as unknowns:

Aldehydes

(100 mL)	2-Butenal (crotonaldehyde)
(100 mL)	Octanal (caprylaldehyde)
(100 mL)	Pentanal (valeraldehyde)

Ketones

(100 mL)	Acetophenone
(100 mL)	Cyclopentanone
(100 mL)	2-Pentanone
(100 mL)	3-Pentanone

Experiment 7   ## Properties of carboxylic acids and esters

Special Equipment

(5 rolls)	pH paper (range 1–12)
(100)	Disposable Pasteur pipets
(5 vials)	Litmus paper, blue
(25)	Hot plates

Chemicals

(10 g)	Salicylic acid
(10 g)	Benzoic acid

The following solutions are placed in dropper bottles.

(75 mL)	Acetic acid
(50 mL)	Formic acid
(25 mL)	Benzyl alcohol
(50 mL)	Ethanol (ethyl alcohol)
(25 mL)	2-Methyl-1-propanol (isobutyl alcohol)
(25 mL)	3-Methyl-1-butanol (isopentyl alcohol)
(50 mL)	Methanol (methyl alcohol)
(25 mL)	Methyl salicylate
(250 mL)	6 M hydrochloric acid, 6 M HCl: take 125 mL of concentrated HCl (12 M HCl) and add to 50 mL of ice cold water; dilute with enough water to 250 mL. **Wear a face shield, rubber gloves, and a rubber apron during the preparation. Do in the hood.**

(100 mL)	3 M hydrochloric acid, 3 M HCL: take 50 mL 6 M HCl and bring to 100 mL; **follow the same precautions as above.**
(300 mL)	6 M sodium hydroxide, 6 M NaOH: dissolve 72.00 g NaOH in enough water to bring to 300 mL; **follow the same precautions as above.**
(150 mL)	2 M sodium hydroxide, 2 M NaOH: take 50 mL 6 M NaOH and bring to 150 mL; **follow the same precautions as above.**
(25 mL)	Concentrated sulfuric acid (18 M H_2SO_4).

Experiment 8   Properties of amines and amides

Special Equipment

(2 rolls)	pH paper (range 0 to 12)
(25)	Hot plates

Chemicals

(20 g)	Acetamide

The following chemicals and solutions should be placed in dropper bottles.

(25 mL)	Triethylamine
(25 mL)	Aniline
(25 mL)	N,N-Dimethylaniline
(100 mL)	Diethyl ether (ether)
(100 mL)	6 M ammonia solution, 6 M NH_3: add 40 mL concentrated NH_3 (28%) to 50 mL water; then add enough water to 100 mL volume. **Do in the hood.**
(100 mL)	6 M hydrochloric acid, 6 M HCl: add 50 mL concentrated HCl (12 M HCl) to 40 mL ice cold water; then add enough water to 100 mL volume. **Wear a face shield, rubber gloves and a rubber apron when preparing. Do in the hood.**
(50 mL)	Concentrated hydrochloric acid (12 M HCl)
(250 mL)	6 M sulfuric acid, 6 M H_2SO_4: pour 83.4 mL concentrated H_2SO_4 (18 M H_2SO_4) into 125 mL ice cold water. Stir slowly. Then add enough water to 250 mL volume. **Wear a face shield, rubber gloves, and a rubber apron when preparing. Do in the hood.**
(250 mL)	6 M sodium hydroxide, 6 M NaOH: dissolve 60.00 g NaOH in 100 mL water. Then add enough water to 250 mL volume. **Do in the hood.**

Experiment 9   Polymerization reactions

Special Equipment

(25)	Hot plates
(25)	Cylindrical paper rolls or sticks
(25)	Bent wire approximately 10 cm long

(25)	10-mL pipets or syringes
(25)	Spectroline pipet fillers
(25)	Beaker tongs

Chemicals

The following chemicals and solutions should be placed in dropper bottles.

(75 mL)	Styrene
(250 mL)	Xylene
(10 mL)	*t*-butyl peroxide benzoate (also called *t*-butyl benzoyl peroxide); store at 4°C.
(75 mL)	20% sodium hydroxide: dissolve 15.00 g NaOH in enough water to make 75 mL
(300 mL)	5% adipoyl chloride: dissolve 15.00 g adipoyl chloride in enough cylohexane to make 300 mL
(300 mL)	5% hexamethylene diamine: dissolve 15.00 g hexamethylene diamine in enough water to make 300 mL
(200 mL)	80% formic acid: add 40 mL water to 160 mL formic acid

Experiment 10 — Preparation of acetylsalicylic acid (aspirin)

Special Equipment

(25)	Büchner funnels (85 mm OD)
(25)	Filtervac or no. 2 neoprene adapters
(1 box)	Filter paper (7.0 cm, Whatman no. 2)
(25)	250-mL filter flasks
(25)	Hot plates

Chemicals

(1 jar)	Boiling chips
(25)	Commercial aspirin tablets
(100 mL)	Concentrated phosphoric acid (15 M H_3PO_4) (in a dropper bottle)
(100 mL)	1% iron(III) chloride: dissolve 1 g $FeCl_3 \cdot 6H_2O$ in enough distilled water to make 100 mL (in a dropper bottle)
(100 mL)	Acetic anhydride, freshly opened bottle
(300 mL)	95% ethanol
(100 g)	Salicylic acid

Experiment 11 — Measurement of the active ingredient in aspirin pills

Special Equipment

(1)	Drying oven at 110°C
(25)	Mortars, 100-mL capacity
(25)	Pestles
(1 box)	Filter paper (7.0 cm, Whatman no. 2)
(1 box)	Microscope slides, 3 × 1 in., plain
(25)	25-mL beakers

Chemicals

(1.5 L)	95% ethanol
(300 g)	Commercial asprin tablets
(100 mL)	Hanus iodine solution: dissolve 1.2 g KI in 80 mL water. Add 0.25 g I_2. Stir until the iodine dissolves. Add enough water to make 100 mL volume. Store in dark dropper bottle.

Experiment 12 Isolation of caffeine from tea leaves

Special Equipment

(25)	Cold finger condensers (115 mm long × 15 mm OD)
(1 box)	Filter paper; 7.0 cm, fast flow (Whatman no.1)
(25)	Hot plates
(50)	Latex tubing, 2-ft. lengths
(1 vial)	Melting point capillaries
(25)	No. 2 neoprene adaptors
(25)	Rubber stopper (no. 6, 1-hole) with glass tubing inserted (10 cm length × 7 mm OD)
(25)	125-mL separatory funnels
(25)	25-mL side-arm filter flasks
(25)	250-mL side-arm filter flasks
(25)	Small sample vials
(1)	Stapler
(50)	Vacuum tubing, 2-ft. lengths
(1 box)	Weighing paper

Chemicals

(1 jar)	Boiling chips
(500 mL)	Dichloromethane, CH_2Cl_2
(25 g)	Sodium sulfate, anhydrous, Na_2SO_4
(50 g)	Sodium carbonate, anhydrous, Na_2CO_3
(50)	Tea bags

Experiment 13 Carbohydrates

Special Equipment

(50)	Medicine droppers
(125)	Microtest tubes or 25 depressions white spot plates
(2 rolls)	Litmus paper, red

Chemicals

(20 g)	Boiling chips
(400 mL)	Fehling's reagent (solutions A and B, from Fisher Scientific Co.)
(200 mL)	3 M NaOH: dissolve 24.00 g NaOH in 100 mL water and then add enough water to 200 mL volume
(200 mL)	2% starch solution: place 4 g soluble starch in a beaker. With vigorous stirring, add 10 mL water to form a thin paste. Boil 190 mL water in another beaker. Add the starch paste to the boiling water and stir until the solution becomes clear. Store in a dropper bottle.

(200 mL)	2% sucrose: dissolve 4 g sucrose in 200 mL water
(50 mL)	3 M sulfuric acid: add 8.5 mL concentrated H_2SO_4 (18 M H_2SO_4) to 30 mL ice cold water; **pour the sulfuric acid slowly along the walls of the beaker, this way it will settle on the bottom without much mixing;** stir slowly in order not to generate too much heat; when fully mixed bring the volume to 50 mL. **Wear a face shield, rubber gloves, and a rubber apron when preparing. Do in the hood.**
(100 mL)	2% fructose: dissolve 2 g fructose in 100 mL water. Store in a dropper bottle.
(100 mL)	2% glucose: dissolve 2 g glucose in 100 mL water. Store in a dropper bottle.
(100 mL)	2% lactose: dissolve 2 g lactose in 100 mL water. Store in a dropper bottle.
(100 mL)	0.01 M iodine in KI: dissolve 1.2 g KI in 80 mL water. Add 0.25 g I_2. Stir until the iodine dissolves. Dilute the solution to 100 mL volume. Store in a dark dropper bottle.

Experiment 14   Preparation and properties of a soap

Special Equipment

(25)	Büchner funnels (85 mm OD)
(25)	No. 7 one-hole rubber stoppers
(1 box)	Filter paper (7.0 cm, Whatman no. 2)
(1 roll)	pHydrion paper (pH range 0 to 12)

Chemicals

(1 jar)	Boiling chips
(1 L)	95% ethanol
(1 L)	Saturated sodium chloride (sat. NaCl): dissolve 360 g NaCl in 1 L water
(1 L)	25% sodium hydroxide (25% NaOH): dissolve 250 g NaOH in 1 L water
(1 L)	Vegetable oil
(100 mL)	5% iron(III) chloride (5% $FeCl_3$): dissolve 5 g $FeCl_3 \cdot 6H_2O$ in 100 mL water. Store in a dropper bottle.
(100 mL)	5% calcium chloride (5% $CaCl_2$): dissolve 5 g $CaCl_2 \cdot H_2O$ in 100 mL water. Store in a dropper bottle.
(100 mL)	Mineral oil. Store in a dropper bottle.
(100 mL)	5% magnesium chloride (5% $MgCl_2$): dissolve 5 g $MgCl_2$ in 100 mL water. Store in a dropper bottle.

Experiment 15   Preparation of a hand cream

Special Equipment

| (25) | Bunsen burners |

Chemicals

| (100 mL) | Triethanolamine |
| (40 mL) | Propylene glycol (1,2-propanediol) |

(500 g)	Stearic acid
(40 g)	Methyl stearate (ethyl stearate may be substituted)
(400 g)	Lanolin
(400 g)	Mineral oil

Experiment 16 Extraction and identification of fatty acids from corn oil

Special Equipment

(12)	Water baths
(2)	Heat lamps or hair dryers
(25)	15×6.5 cm silica gel TLC plates
(25)	Rulers, metric scale
(25)	Polyethylene, surgical gloves
(150)	Capillary tubes, open on both ends
(1)	Drying oven, 110°C

Chemicals

(50 g)	Corn oil
(5 mL)	Methyl palmitate solution: dissolve 25 mg methyl palmitate in 5 mL petroleum ether
(5 mL)	Methyl oleate solution: dissolve 25 mg methyl oleate in 5 mL petroleum ether
(5 mL)	Methyl linoleate solution: dissolve 25 mg methyl linoleate in 5 mL petroleum ether
(100 mL)	0.5 M KOH: dissolve 2.81 g KOH in 25 mL water and add 75 ml of 95% ethanol
(500 g)	Sodium sulfate, Na_2SO_4, anhydrous, granular
(100 mL)	Concentrated hydrochloric acid (12 M HCl)
(1 L)	Petroleum ether (b.p. 30–60°C)
(300 mL)	Methanol: perchloric acid mixture: mix 285 mL methanol with 15 mL $HClO_4 \cdot 2H_2O$ (73% perchloric acid)
(400 mL)	Hexane:diethyl ether mixture: mix 320 mL hexane with 80 mL diethyl ether
(10 g)	Iodine crystals, I_2

Experiment 17 Analysis of lipids

Special Equipment

| (25) | Hot plates |
| (25) | Cheesecloth 3×3 in. |

Chemicals

(3 g)	Cholesterol (ash free) 95–98% pure from Sigma Co.
(3 g)	Lecithin (prepared from dried egg yolk) 60% pure from Sigma Co.
(10 g)	Glycerol
(10 g)	Corn oil
(10 g)	Butter
(1)	Egg yolk obtained from one fresh egg before the lab period. Stir and mix.

(250 mL)	Molybdate solution: dissolve 0.8 g $(NH_4)_6Mo_7O_{24} \cdot 4H_2O$ in 30 mL water. Put in an ice bath. Pour slowly 20 mL concentrated sulfuric acid (18 M H_2SO_4) into the solution and stir slowly. After cooling to room temperature bring the volume to 250 mL. **Wear a face shield, rubber gloves, and a rubber apron during the preparation. Do in the hood.**
(50 mL)	0.1 M ascorbic acid solution: dissolve 0.88 g ascorbic acid (vitamin C) in water and bring it to 50 mL volume. This must be prepared fresh every week and stored at 4°C.
(250 mL)	6 M sodium hydroxide, 6 M NaOH: dissolve 60.00 g NaOH in water and bring the volume to 250 mL
(250 mL)	6 M nitric acid, 6 M HNO_3: into a 250-mL volumetric flask containing 100 mL ice cold water, pipet 125 mL concentrated nitric acid (12 M HNO_3); add enough water to bring to 250 mL. **Wear a face shield, rubber gloves, and a rubber apron during the preparation. Do in the hood.**
(200 mL)	Chloroform
(75 mL)	Acetic anhydride
(50 mL)	Concentrated sulfuric acid (18 M H_2SO_4)
(75 g)	Potassium hydrogen sulfate, $KHSO_4$

Experiment 18   TLC separation of amino acids

Special Equipment

(1)	Drying oven, 105–110°C
(2)	Heat lamps or hair dryers
(50)	15 × 6.5 cm silica gel TLC plates (or chromatographic paper Whatman no. 1)
(25)	Rulers, metric scale
(25)	Polyethylene, surgical gloves
(150)	Capillary tubes, open on both ends
(1 roll)	Aluminum foil
(2)	Wide-mouth jars

Chemicals

(25 mL)	0.12% aspartic acid solution: dissolve 30 mg aspartic acid in 25 mL distilled water
(25 mL)	0.12% phenylalanine solution: dissolve 30 mg phenylalanine in 25 mL distilled water
(25 mL)	0.12% leucine solution: dissolve 30 mg leucine in 25 mL distilled water
(25 mL)	Aspartame solution: dissolve 150 mg Equal sweetener powder in 25 mL distilled water
(50 mL)	3 M HCl solution: place 10 mL ice cold distilled water into a 50-mL volumetric flask. Add slowly 12.5 mL of concentrated HCl (12 M HCl) and bring it to volume with distilled water. **Wear a face shield, rubber gloves, and a rubber apron when preparing. Do in the hood.**

(1 L)	Solvent mixture: mix 600 mL 1-butanol with 150 mL acetic acid and 250 mL distilled water
(1 can)	Ninhydrin spray reagent (0.2% ninhydrin in ethanol or acetone). Do not use any reagent older than 6 months.
(1 can)	Diet Coke
(4 packets)	Equal or NutraSweet, sweeteners
(10 g)	Iodine crystals, I_2

Experiment 19 Acid–base properties of amino acids

Special Equipment

(10)	pH meters or
(5 rolls)	pHydrion short-range papers, from each range: pH: 0.0 to 3.0; 3.0 to 5.5; 5.2 to 6.6; 6.0 to 8.0; 8.0 to 9.5 and 9.0 to 12.0
(25)	20-mL pipets
(25)	50-mL burets
(25)	Spectroline pipet fillers
(25)	Pasteur pipets

Chemicals

| (500 mL) | 0.25 M NaOH: dissolve 5.00 g NaOH in 100 mL water and then add enough water to 500 mL volume |
| (750 mL) | 0.1 M alanine solution: dissolve 6.68 g L-alanine in 500 mL; add sufficient 1 M HCl to bring the pH to 1.5. Add enough water to 750 mL volume. |

or

Do as above but use either 5.63 g glycine or 9.84 g leucine or 12.39 g phenylalanine or 8.79 g valine.

Experiment 20 Isolation and identification of casein

Special Equipment

(25)	Hot plates
(25)	600-mL beakers
(25)	Büchner funnels (O.D. 85 mm) in no. 7 1-hole rubber stopper
(7 boxes)	Whatman no. 2 filter paper, 7 cm
(25)	Rubber bands
(25)	Cheese cloths (6 × 6 in.)

Chemicals

(1 jar)	Boiling chips
(1 L)	95% ethanol
(1 L)	Diethyl ether:ethanol mixture (1:1)
(0.5 gal)	Regular milk
(500 mL)	Glacial acetic acid

The following solutions should be placed in dropper bottles:

| (100 mL) | Concentrated nitric acid (12 M HNO_3) |
| (100 mL) | 2% albumin suspension: dissolve 2 g albumin in 100 mL water |

(100 mL)	2% gelatin: dissolve 2 g gelatin in 100 mL water
(100 mL)	2% glycine: dissolve 2 g glycine in 100 mL water
(100 mL)	5% copper(II) sulfate: dissolve 5 g $CuSO_4$ (or 7.85 g $CuSO_4 \cdot 5H_2O$) in 100 mL water
(100 mL)	5% lead(II) nitrate: dissolve 5 g $Pb(NO_3)_2$ in 100 mL water
(100 mL)	5% mercury(II) nitrate: dissolve 5 g $Hg(NO_3)_2$ in 100 mL water
(100 mL)	Ninhydrin reagent: dissolve 3 g ninhydrin in 100 mL acetone. Do not use a reagent older than 6 months.
(100 mL)	10% sodium hydroxide: dissolve 10 g NaOH in 100 mL water
(100 mL)	1% tyrosine: dissolve 1 g tyrosine in 100 mL water
(100 mL)	5% sodium nitrate: dissolve 5 g $NaNO_3$ in 100 mL water

Experiment 21   Isolation and identification of DNA from yeast

Special Equipment

(12)	Mortars
(12)	Pestles
(6)	Desk top clinical centrifuges (swinging bucket rotor) (optional)

Chemicals

(100 g)	Baker's yeast, freshly purchased
(500 g)	Acid-washed sand
(1 L)	Saline-CTAB isolation buffer: dissolve 20 g hexadecyltrimethylammonium bromide (CTAB, Sigma 45882), 2 mL 2-mercaptoethanol, 7.44 g ethylenediamine tetraacetate (EDTA, Sigma ED2SS), 8.77 g NaCl in 1 L Tris buffer. The Tris buffer is prepared by dissolving 12.1 g Tris in 700 mL water; adjust the pH to 8 by titrating with 4 M HCl. Add enough water to bring the volume to 1 L.
(200 mL)	6 M sodium perchlorate solution, 6 M $NaClO_4$: dissolve 147 g $NaClO_4$ in 100 mL water and add enough water to bring the volume to 200 mL
(100 mL)	Citrate buffer: dissolve 0.88 g NaCl and 0.39 g sodium citrate in 100 mL water
(1 L)	Chloroform-isopentyl alcohol mixture: to 960 mL chloroform, add 40 mL isopentyl alcohol. Mix throughly.
(2 L)	2-Propanol (isopropyl alcohol)
(50 mL)	1% glucose solution: dissolve 0.5 g D-glucose in 50 mL water
(50 mL)	1% ribose solution: dissolve 0.5 g D-ribose in 50 mL water
(50 mL)	1% deoxyribose solution: dissolve 0.5 g 2-deoxy-D-ribose in 50 mL water
(200 mL)	95% ethanol
(500 mL)	Diphenylamine reagent. *This must be prepared shortly before lab use.* Dissolve 7.5 g reagent grade diphenylamine (Sigma D3409) in 50 mL glacial acetic acid. Add 7.5 mL concentrated sulfuric acid (18 M H_2SO_4). Prior to use add

2.5 mL 1.6% acetaldehyde (made by dissolving 0.16 g acetaldehyde in 10 mL water). **Wear a face shield, rubber gloves, and a rubber apron when preparing. Do in the hood.**

Experiment 22   Viscosity and secondary structure of DNA

Equipment

(5)	Ostwald (or Cannon-Ubbelhode) capillary viscometers; 3-mL capacity, approximate capillary diameter 0.2 mm; **efflux time of water = 40–50 sec.**
(5)	Stopwatches (Wristwatches can also time the efflux with sufficient precision.)
(5)	Stands with utility clamps
(25)	Pasteur pipets
(10)	Spectroline pipet fillers

Chemicals

(500 mL)	Buffer solution: dissolve 4.4 g sodium chloride, NaCl, and 2.2 g sodium citrate, $Na_3C_6H_5O_7 \cdot 2H_2O$ in 450 mL distilled water. Adjust the pH with either 0.1 M HCl or 0.1 M NaOH to pH 7.0. Add enough water to bring to 500 mL volume.
(200 mL)	DNA solution: dissolve 20 mg of calf thymus Type I highly polymerized DNA (obtainable from Sigma as well as from other companies) in 200-mL buffer solution at pH 7.0. The purchased DNA powder should be kept in the freezer. The DNA solutions should be prepared fresh or maximum 2–3 hr. in advance of the experiment. The solution should be kept at 4°C; 1–2 hr. before the experiment, the solution should be allowed to come to room temperature. Label the solution as 0.01 g/dL concentration.
(100 mL)	1 M hydrochloric acid, 1 M HCl: add 8.3 mL concentrated HCl (12 M HCl) to 50 mL ice cold water; add enough water to bring to 100 mL volume. **Wear a face shield, rubber gloves, and a rubber apron during preparation. Do in the hood.**
(100 mL)	0.1 M hydrochloric acid: add 10.0 mL 1 M HCl to 50 mL water; add enough water bring to 100 mL volume. **Follow safety procedure described above.**
(100 mL)	1 M sodium hydroxide: dissolve 4.00 g NaOH in 50 mL water; add enough water to bring to 100 mL volume.
(100 mL)	0.1 M sodium hydroxide: dissolve 0.40 g NaOH in 50 mL water; add enough water to bring to 100 mL volume.

Experiment 23   Kinetics of urease catalyzed decomposition of urea

Special Equipment

(25)	5-mL pipets
(25)	10-mL graduated pipets
(25)	10-mL volumetric pipets

(25)	50-mL burets
(25)	Buret holders
(25)	Spectroline pipet fillers

Chemicals

(3.5 L)	0.05 M Tris buffer: dissolve 21.05 g Tris buffer in water (3 L). Adjust the pH to 7.2 with 1 M HCl solution; add sufficient water to make 3.5 L. Portions of buffer solution will be used to make urea and enzyme solutions.
(2.5 L)	0.3 M urea solution: dissolve 45 g urea in 2.5 L Tris buffer
(50 mL)	1×10^{-3} M phenylmercuric acetate: dissolve 16.5 mg phenylmercuric acetate in 40 mL water; add enough water to bring the volume to 50 mL.
	CAUTION! Phenylmercuric acetate is a poison. Do not touch the chemical with your hands. Do not swallow the solution. Wear rubber gloves in the preparation.
(50 mL)	1% $HgCl_2$ solution: dissolve 0.5 g $HgCl_2$ in enough water to make 50 mL solution
(100 mL)	0.04% methyl red indicator: dissolve 40 mg methyl red in 100 mL distilled water
(500 mL)	Urease solution: prepare the enzyme solution on the week of the experiment and store at 4°C. Take 1.0 g urease, dissolve in 500 mL Tris buffer. (One can buy urease with 5 to 6 units activity, for example, from Nutritional Biochemicals, Cleveland, Ohio.) The activity of the enzyme printed on the label should be checked by the stockroom personnel or instructor.
(1.0 L)	0.05 N HCl: add 4.2 mL concentrated HCl (12 M HCl) to 100 mL ice cold water; add enough water to bring to 1.0 L volume. **Wear a face shield, rubber gloves, and a rubber apron during the preparation. Do in the hood.**

Experiment 24   Isocitrate dehydrogenase—an enzyme of the citric acid cycle

Special Equipment

| (15) | Spectrophotometers with 5 cuvettes each |
| (25) | 1-mL graduated pipets |

Chemicals

| (40 mL) | Phosphate buffer at pH 7.0: mix together 25 mL 0.1 M KH_2PO_4 and 15 mL 0.1 M NaOH. To prepare 0.1 M NaOH, add 0.20 g NaOH to 20 mL water in a 50-mL volumetric flask; stir to dissolve; add enough water to bring to 50 mL volume. To prepare 0.1 M KH_2PO_4, add 0.68 g potassium dihydrogen phosphate to 40 mL water in a 50-mL volumetric flask; stir to dissolve; add enough water to bring to 50 mL volume. |
| (20 mL) | 0.1 M $MgCl_2$: add 0.19 g magnesium chloride to 20 mL water; stir to dissolve. |

(50 mL)	Isocitrate dehydrogenase: commercial preparations from porcine heart are obtainable from companies such as Sigma, etc. (EC 1.1.1.42) (activity about 8 units per mg of solid). Dissolve 10 mg of the enzyme in 50 mL water. This solution should be made fresh before the lab period and kept in a refrigerator until used.
(20 mL)	6.0 mM β-Nicotinamide adenine dinucleotide, β-NADP$^+$, solution: dissolve 92 mg NADP$^+$ in 20 mL water
(50 mL)	15 mM sodium isocitrate solution: dissolve 160 mg sodium isocitrate in 50 mL water

Experiment 25   **Quantitative analysis of vitamin C contained in foods**

Special Equipment

(25)	50-mL burets
(25)	Buret clamps
(25)	Ring stands
(25)	Spectroline pipet fillers
(25)	10-mL volumetric pipets
(1 box)	Cotton

Chemicals

(500 g)	Celite, filter aid
(1 can)	Hi-C orange drink
(1 can)	Hi-C grapefruit drink
(1 can)	Hi-C apple drink
(2 L)	0.01 M iodine solution: add 32 g KI to 800 mL water; stir to dissolve. Add 5 g I$_2$; stir to dissolve. Add enough water to bring to 2 L volume. Store in dark bottle. **Caution! Iodine is poisonous if taken internally.**
(100 mL)	3 M HCl: add 25 mL concentrated HCl (12 M HCl) to 50 mL ice cold water; add enough water to bring to 100 mL volume. **Wear a face shield, rubber gloves, and a rubber apron during the preparation. Do in the hood.**
(100 mL)	2% starch solution: place 2 g soluble starch in a 50-mL beaker. Add 10 mL water. Stir vigorously to form a paste. Boil 90 mL water in a second beaker. Add the starch paste to the boiling water. Stir until the solution becomes translucent. Cool to room temperature.

Experiment 26   **Analysis of vitamin A in margarine**

Special Equipment

(1)	UV spectrophotometer with suitable UV light source. Preferably it should be able to read down to 200 nm.
(1 pair)	Matched quartz cells, with 1-cm internal path length.
(2)	Long-wavelength UV lamp. The lamp should provide radiation in the 300-nm range (for example, #UVSL-55; LW 240 from Ultraviolet Products Inc.)

(12)	500-mL separatory funnels
(12)	25-mL (or 50-mL) burets
(12)	Hot plates, each with a water bath
(12)	Beaker tongs

Chemicals

(0.5 lb)	Margarine
(400 mL)	50% KOH; weigh 200 g KOH and add 200 mL water, with constant stirring; add enough water to make 400 mL
(1 L)	95% ethanol
(150 mL)	Absolute ethanol
(2 L)	Petroleum ether, 30–60°C
(3 L)	Diethyl ether
(350 g)	Alkaline aluminum oxide (alumina)

Experiment 27   ## Urine analysis

Special Equipment

(12)	Hydrometers (urinometers) from 1.00 to 1.40 specific gravity
(3 bottles)	Clinistix (50 reagent strips/bottle)
(3 bottles)	Urobilistix (50 reagent strips/bottle)
(3 bottles)	Phenistix (50 reagent strips/bottle)
(1 bottle)	Albustix (100 reagent strips/bottle)
(1 bottle)	Ketostix (100 reagent strip/bottle)

These are obtainable from Ames Co., Division Miles Lab. Inc., Elkhart, Indiana, 46515. Instead of the individual "Stix," you may purchase 4 bottles of multipurpose Labstix (100 reagent strips/bottle).

Chemicals

(500 mL)	Normal urine. This and all other urine samples must be kept at 4°C until 30 min. prior to the lab period. Alternatively, you may ask each student to provide fresh urine samples for analysis.
(500 mL)	"Pathological urine A": add 4 g glucose, 2 mL acetone and 2 g citric acid to 500 mL water
(500 mL)	"Pathological urine B": add 50 mg phenylpyruvate, and 500 mg sodium phoshate, Na_3PO_4, to 500 mL water
(250 mL)	1% glucose: dissolve 2.5 g glucose in 250 mL water
(200 mL)	0.25% glucose: dilute 50 mL of 1% glucose solution with water to 200 mL volume